Everyday Mathematics®

The University of Chicago School Mathematics Project

Student Math Journal
Volume 2

The McGraw·Hill Companies

The University of Chicago School Mathematics Project (UCSMP)

Max Bell, Director, UCSMP Elementary Materials Component; Director, *Everyday Mathematics* First Edition; James McBride, Director, *Everyday Mathematics* Second Edition; Andy Isaacs, Director, *Everyday Mathematics* Third Edition; Amy Dillard, Associate Director, *Everyday Mathematics* Third Edition

Authors

Max Bell, John Bretzlauf, Amy Dillard, Robert Hartfield, Andy Isaacs, James McBride, Ann McCarty*, Kathleen Pitvorec, Peter Saecker, Robert Balfanz†, William Carroll†

**Third Edition only* *†First Edition only*

Technical Art

Diana Barrie

Teacher in Residence

Denise Porter

Photo Credits

Cover (l)Stuart Westmoreland/CORBIS, (c)Digital Vision/Getty Images, (r)Kelly Kalhoefer/Getty Images; **Back Cover** Digital Vision/Getty Images; **iii** The McGraw-Hill Companies; **iv** (t)Dana Tezarr/Getty Images, (b)The McGraw-Hill Companies; **v** (t)Getty Images, (b)Royalty-Free/CORBIS; **vi** Royalty-Free/CORBIS; **vii** Martin Barraud/Getty Images; **viii** Adam Crowley/Getty Images; **233 267 322** The McGraw-Hill Companies.

Contributors

Ann Brown, Sarah Busse, Terry DeJong, Craig Dezell, John Dini, James Flanders, Donna Goffron, Steve Heckley, Karen Hedberg, Deborah Arron Leslie, Sharon McHugh, Janet M. Meyers, Donna Owen, William D. Pattison, Marilyn Pavlak, Jane Picken, Kelly Porto, John Sabol, Rose Ann Simpson, Debbi Suhajda, Laura Sunseri, Jayme Tighe, Andrea Tyrance, Kim Van Haitsma, Mary Wilson, Nancy Wilson, Jackie Winston, Carl Zmola, Theresa Zmola

This material is based upon work supported by the National Science Foundation under Grant No. ESI-9252984. Any opinions, findings, conclusions, or recommendations expressed in this material are those of the authors and do not necessarily reflect the views of the National Science Foundation.

www.WrightGroup.com

Printed in the United States of America.

Send all inquiries to:
Wright Group/McGraw-Hill
P.O. Box 812960
Chicago, IL 60681

ISBN 0-07-605274-5

17 18 19 QDB 13 12 11

The McGraw·Hill Companies

Contents

UNIT 6 Number Systems and Algebra Concepts

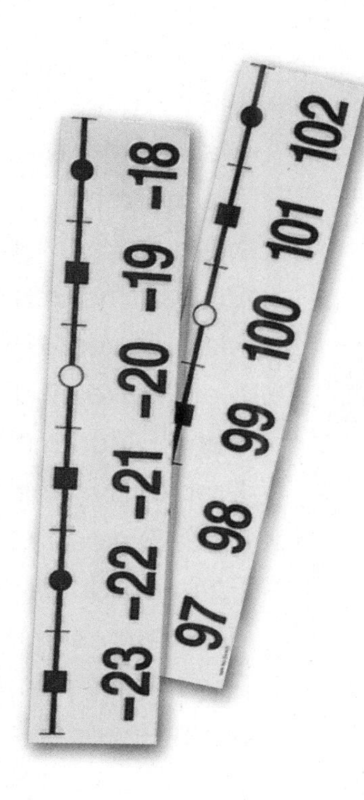

UNIT 7 Probability and Discrete Mathematics

UNIT 8 Rates and Ratios

UNIT 9 More about Variables, Formulas, and Graphs

UNIT 10 Geometry Topics

References

Activity Sheets

Date Time

LESSON 6•1

Applying Properties of Multiplication

Math Message

1. Write a general pattern in words for the group of three special cases.

$19 * 1 = 19$

$\frac{2}{7} * 1 = \frac{2}{7}$

$0.084 * 1 = 0.084$

General pattern: ______________________________

2. Write a general pattern in words for the group of three special cases.

$\frac{58}{58} = 1$

$\frac{\frac{3}{8}}{\frac{3}{8}} = 1$

$\frac{7.02}{7.02} = 1$

General pattern: ______________________________

3. Multiply. Write your answers in simplest form.

a. $4 * \frac{19}{19} =$ ______ **b.** $\frac{2}{3} * \frac{6}{6} =$ ______ **c.** $0.5 * \frac{2}{2} =$ ______

Multiply. Write your answers in simplest form. When you and your partner have finished solving the problems, compare your answers.

4. $\frac{5}{6} * \frac{3}{10} =$ ______
5. $6 * \frac{2}{3} =$ ______
6. $7 * \frac{3}{7} =$ ______
7. $2\frac{3}{4} * \frac{4}{1} =$ ______
8. $2\frac{3}{5} * 1\frac{2}{3} =$ ______
9. $\frac{7}{3} * \frac{1}{3} =$ ______
10. $\frac{1}{4} * \frac{2}{5} =$ ______
11. $3\frac{3}{8} * \frac{3}{4} =$ ______
12. $1\frac{5}{6} * 4\frac{2}{3} =$ ______
13. $\frac{7}{10} * 2\frac{3}{5} =$ ______
14. $\frac{4}{1} * \frac{1}{4} =$ ______
15. $\frac{1}{100} * \frac{100}{1} =$ ______
16. $\frac{7}{8} * \frac{8}{7} =$ ______
17. $1\frac{5}{6} * \frac{6}{11} =$ ______
18. Write three special cases for the general pattern $x * \frac{1}{x} = 1$.

______________ ______________ ______________

Date _______ Time _______

LESSON 6•1

Reciprocals

Reciprocal Property

If a and b are any numbers except 0, then $\frac{a}{b} * \frac{b}{a} = 1$.

$\frac{a}{b}$ and $\frac{b}{a}$ are called reciprocals of each other.

$a * \frac{1}{a} = 1$, so a and $\frac{1}{a}$ are reciprocals of each other.

Find the reciprocal of each number. Multiply to check your answers.

1. 6 _______
2. 17 _______
3. $\frac{3}{4}$ _______
4. $\frac{1}{3}$ _______
5. $\frac{3}{8}$ _______
6. $\frac{13}{16}$ _______
7. $8\frac{1}{2}$ _______
8. $3\frac{5}{6}$ _______
9. $4\frac{2}{3}$ _______
10. $6\frac{1}{4}$ _______
11. 0.1 _______
12. 0.4 _______
13. 0.75 _______
14. 2.5 _______
15. 0.375 _______
16. 5.6 _______

Solve mentally.

17. $3\frac{1}{2} * 4 * \frac{1}{4}$ _______
18. $\frac{1}{6} * \frac{2}{5} * 6$ _______
19. $\frac{5}{7} * \frac{7}{5} * 4\frac{5}{7}$ _______
20. $2 * 8\frac{1}{2} * \frac{1}{2}$ _______
21. $3\frac{1}{3} * \frac{3}{10} * \frac{7}{10} * 1\frac{3}{7}$ _______
22. $3.875 * 2.5 * 0.4$ _______

Date Time

LESSON 6•1

Math Boxes

1. Rename each mixed number as a fraction.

a. $3\frac{7}{8}$ = ________

b. ________ = $5\frac{8}{9}$

c. ________ = $8\frac{5}{6}$

d. ________ = $6\frac{9}{7}$

e. $14\frac{2}{3}$ = ________

2. Multiply.

a. $3\frac{1}{2} * 4 =$ ________

b. $\frac{1}{2} * 2\frac{1}{3} =$ ________

c. $\frac{1}{8} * \frac{2}{9} * 8 =$ ________

d. ________ $= \frac{1}{5} * \frac{1}{2} * 10$

3. Circle the number sentence that describes the numbers in the table.

A. $y = x + 10$

B. $(2 * x) + 5 = y$

C. $y - 2 = (5 - x)$

D. $y - 8 = x$

x	*y*
3	11
5	15
0	5
10	25

4. Write each number in standard notation.

a. $(5 * 10^1) + (3 * 10^0) + (4 * 10^{-2})$

____

b. $(9 * \frac{1}{10}) + (7 * \frac{1}{100}) + (6 * \frac{1}{1{,}000})$

____

5. Write a percent for each fraction.

a. $\frac{4}{5}$ = ________

b. $\frac{1}{8}$ = ________

c. $\frac{7}{8}$ = ________

d. $1\frac{3}{4}$ = ________

e. $\frac{3}{100}$ = ________

6. a. Use your Geometry Template to draw a spinner with colored sectors so the chances of landing on these colors are as follows:

red: $\frac{3}{10}$

blue: 0.33

green: 20%

b. On this spinner, what is the chance of *not* landing on red, blue, or green? ________

Date Time

LESSON 6•2

Dividing Fractions and Mixed Numbers

Math Message

Solve the problems. Use a ruler to help you.

1. How many 3-centimeter segments are in 12 centimeters? ______________

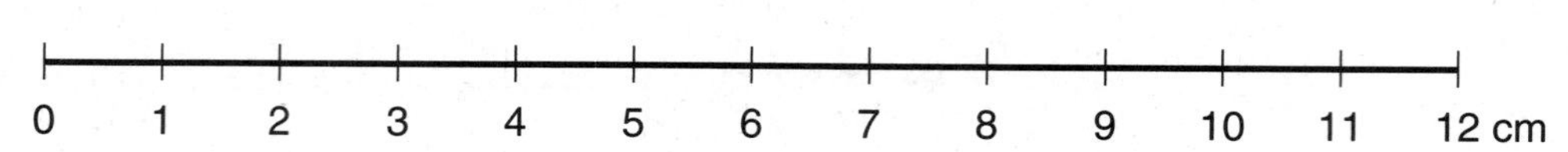

2. How many $\frac{1}{2}$-inch segments are in 4 inches? ______________

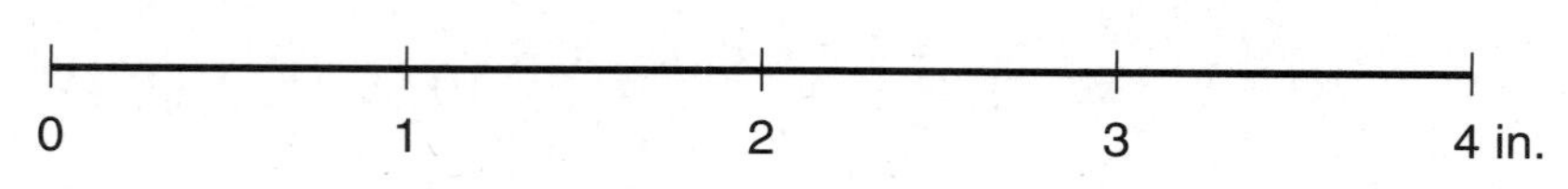

3. How many $\frac{3}{4}$-inch segments are in 3 inches? ______________

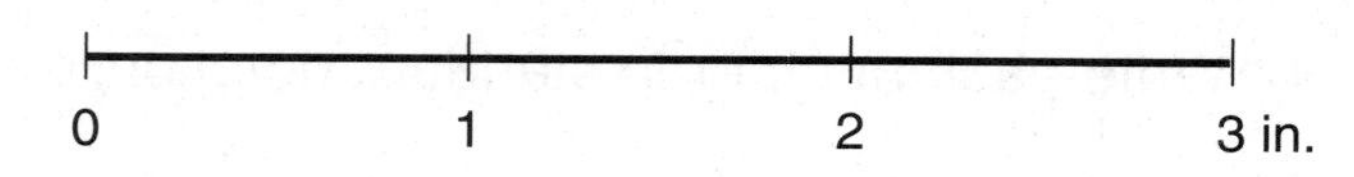

4. How many $\frac{3}{4}$-inch segments are in $4\frac{1}{2}$ inches? ______________

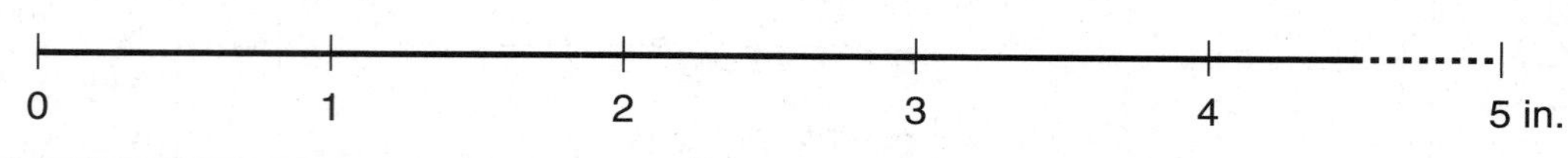

Division of Fractions Algorithm

$$\frac{a}{b} \div \frac{c}{d} = \frac{a}{b} * \frac{d}{c}$$

Divide. Show your work. Write your answers in simplest form.

5. $\frac{3}{8} \div \frac{5}{6} =$ ______________

6. $\frac{4}{7} \div \frac{2}{3} =$ ______________

7. $\frac{3}{10} \div \frac{3}{5} =$ ______________

8. $\frac{11}{12} \div \frac{8}{5} =$ ______________

Date ____________________ Time ____________________

LESSON 6•2

Dividing Fractions and Mixed Numbers *cont.*

Divide. Show your work. Write your answers in simplest form.

9. $\frac{7}{8} \div \frac{4}{9} =$ ____________________

10. $\frac{7}{12} \div \frac{1}{3} =$ ____________________

11. $\frac{5}{9} \div \frac{1}{10} =$ ____________________

12. $\frac{3}{4} \div \frac{7}{8} =$ ____________________

13. $\frac{5}{3} \div \frac{3}{5} =$ ____________________

14. $\frac{9}{10} \div \frac{2}{3} =$ ____________________

15. $1\frac{5}{8} \div \frac{4}{6} =$ ____________________

16. $\frac{3}{8} \div \frac{8}{2} =$ ____________________

17. $\frac{5}{4} \div \frac{16}{8} =$ ____________________

18. $1\frac{2}{3} \div 2\frac{1}{4} =$ ____________________

19. $\frac{8}{9} \div \frac{8}{9} =$ ____________________

20. $3\frac{7}{8} \div 1\frac{3}{4} =$ ____________________

21. Explain how you found your answer to Problem 19.

____

____

____

22. How is dividing 5 by $\frac{1}{5}$ different from multiplying 5 by $\frac{1}{5}$?

____

____

____

Date Time

LESSON 6•2

Math Boxes

1. Write the reciprocal.

a. $\frac{3}{8}$ ________

b. $\frac{5}{9}$ ________

c. $1\frac{3}{4}$ ________

d. 0.68 ________

2. Divide. Simplify if possible.

a. $8 \div \frac{4}{5} =$ ________

b. $5\frac{1}{5} \div \frac{2}{5} =$ ________

c. ________ $= \frac{2}{9} \div \frac{1}{3}$

d. ________ $= \frac{9}{14} \div \frac{3}{7}$

3. Lines *l* and *m* are parallel. Without using a protractor, find the degree measure of each numbered angle. Write each measure on the drawing.

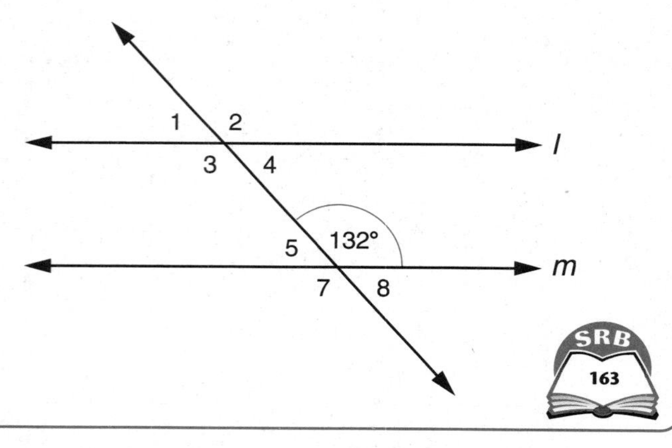

SRB 163

4. There are 30.48 centimeters in 1 foot.

Complete each statement.

a. ____________ mm = 1 ft

b. ____________ cm = 1 yd

c. 304.8 cm = ____________ ft

d. ____________ cm = 1 in.

5. Express each decimal as a percent.

a. 0.82 = __________

b. __________ = 0.4375

c. 0.077 = __________

d. __________ = 0.009

6. If you randomly pick a date in April, how many equally likely outcomes are there?

____

Explain your answer.

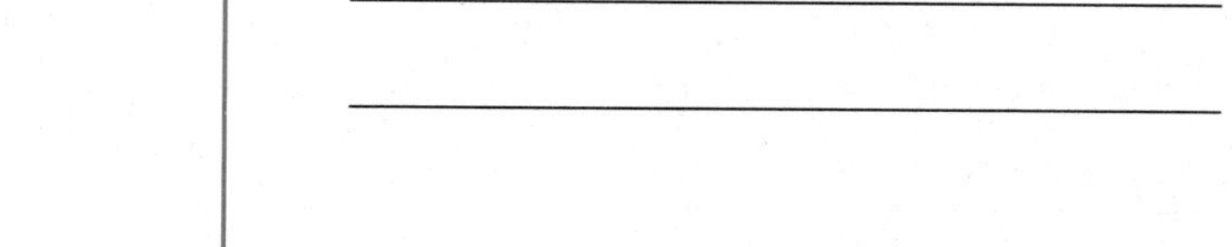

Date _______________ Time _______________

LESSON 6•3 Negative Numbers on a Calculator

Math Message

Read the section "Negative Numbers" on page 271 in your *Student Reference Book.* Study the key sequence for the calculator you are using.

1. Enter each number into your calculator. Record the calculator display.

Enter	-4.85	$-(\frac{2}{3})$	-0.006	$(-4)^2$	-8^5
Display	________	________	________	________	________

You can use the negative sign (−) or OPP to represent the phrase "the opposite of." For example, "the opposite of 12" is written as −12 or OPP(12). In the same way, "the opposite of −5" is written as −(−5) or OPP(−5).

2. Enter the first number into your calculator. Record the calculator display. Clear the calculator before entering the next number.

Enter	OPP(75)	OPP(−89)	OPP$(-3)^2$	OPP(15 − 21)
Display	________	________	________	________

Use a calculator to add or subtract. Remember, the term OPP means "the opposite of."

3. −26 − 17 = ________

   −26 + OPP(17) = ________

4. −34 − 68 = ________

   −34 + (−68) = ________

5. 56 − 24 = ________

   56 + OPP(24) = ________

6. 18 − 84 = ________

   18 + (−84) = ________

7. 43 − (−97) = ________

   43 + OPP(−97) = ________

   43 + 97 = ________

8. 31 − (−13) = ________

   31 + (−(−13)) = ________

   31 + 13 = ________

9. −130 − (−62) = ________

   −130 + OPP(−62) = ________

   −130 + 62 = ________

10. −2 − (−22) = ________

    −2 + (−(−22)) = ________

    −2 + 22 = ________

Date _______________ Time _______________

LESSON 6•3

Subtracting Positive and Negative Numbers

One way to subtract one number from another number is to change the subtraction problem into an addition problem.

Subtraction Rule

To subtract a number b from a number a, add the opposite of b to a. Thus, for any numbers a and b, $a - b = a + \text{OPP}(b)$, or $a - b = a + (-b)$.

Examples:

$6 - 9 = 6 + \text{OPP}(9) = 6 + (-9) = -3$

$-15 - (-23) = -15 + \text{OPP}(-23) = -15 + 23 = 8$

Rewrite each subtraction problem as an addition problem. Then solve the problem.

1. $22 - (15) =$ $22 + \text{OPP}(15) = 7$
2. $-35 - 20 =$ _______________
3. $-3 - (-4.5) =$ _______________
4. $-27 - (-27) =$ _______________

Subtract.

5. $-23 - (-5) =$ ________
6. $9 - (-54) =$ ________
7. $-(\frac{4}{5}) - 1\frac{1}{5} =$ ________
8. $\$1.25 - (-\$6.75) =$ ________
9. $-76 - (-56) =$ ________
10. $-27 - 100 =$ ________
11. Explain how you solved Problem 9. _______________

____

____

Fill in the missing numbers.

12. ________ $+ 5 = -10$

    $-10 - 5 =$ ________
13. ________ $+ (-5) = -10$

    $-10 - (-5) =$ ________
14. $-9 +$ ________ $= 0$

    $0 - (-9) =$ ________
15. $16 +$ ________ $= -7$

    $-7 - 16 =$ ________
16. $-25 +$ ________ $= 15$

    $15 - (-25) =$ ________
17. ________ $+ 13 = -8$

    $-8 - 13 =$ ________

Date Time

LESSON 6•3

Math Boxes

1. Rename each fraction as a mixed number.

a. $\frac{320}{25}$ = ______

b. ______ = $\frac{43}{7}$

c. ______ = $\frac{101}{5}$

d. ______ = $\frac{75}{8}$

e. $\frac{147}{4}$ = ______

2. Multiply.

a. $\frac{1}{3} * 12\frac{2}{9} * 3$ = ______

b. $\frac{4}{5} * \frac{1}{2} * 4$ = ______

c. $6\frac{2}{3} * \frac{1}{5} * \frac{3}{20}$ = ______

d. ______ = $\frac{2}{5} * \frac{3}{8} * \frac{15}{6}$

3. Circle the number sentence that describes the numbers in the table.

A. $p = m * 2$

B. $(3 - m) = p + 8$

C. $p = (3 * m) - 8$

D. $m - 8 = p$

m	*p*
8	16
0	−8
4	4
10	22

4. Which of the following is $(3 * 10^3) + (7 * 10^0) + (5 * 10^{-2})$ written in standard notation? Choose the best answer.

- 30.2
- 30.50
- 307.005
- 3,007.05

5. Write a percent for each fraction.

a. $\frac{10}{50}$ = ______

b. $\frac{2}{3}$ = ______

c. $\frac{24}{25}$ = ______

d. $\frac{11}{8}$ = ______

e. $\frac{2}{1{,}000}$ = ______

6. a. Use your Geometry Template to draw a spinner with colored sectors so the chances of landing on these colors are as follows:

red: 1 out of 4

blue: $\frac{3}{8}$

b. On this spinner, what is the chance of *not* landing on red or blue?

______

Date Time

LESSON 6•4 Multiplication Patterns

In each of Problems 1–4, complete the patterns in Part a. Check your answers with a calculator. Then circle the word in parentheses that correctly completes the statement in Part b.

1. a. $6 * 4 = 24$
$6 * 3 = 18$
$6 * 2 =$ ________
$6 * 1 =$ ________
$6 * 0 =$ ________

b. Positive * Positive Rule:

When a positive number is multiplied by a positive number, the product is a

(positive or negative) number.

2. a. $5 * 2 = 10$
$5 * 1 = 5$
$5 * 0 = 0$
$5 * (-1) =$ ________
$5 * (-2) =$ ________

b. Positive * Negative Rule:

When a positive number is multiplied by a negative number, the product is a

(positive or negative) number.

3. a. $2 * 3 = 6$
$1 * 3 = 3$
$0 * 3 = 0$
$-1 * 3 =$ ________
$-2 * 3 =$ ________

b. Negative * Positive Rule:

When a negative number is multiplied by a positive number, the product is a

(positive or negative) number.

4. a. $-4 * 1 = -4$
$-4 * 0 = 0$
$-4 * (-1) = 4$
$-4 * (-2) =$ ________
$-4 * (-3) =$ ________

b. Negative * Negative Rule:

When a negative number is multiplied by a negative number, the product is a

(positive or negative) number.

5. a. Solve.
$-1 * 6 =$ ________
$-1 * (-7.7) =$ ________
$-1 * -(-\frac{1}{2}) =$ ________
$-1 * m =$ ________

b. Multiplication Property of −1:

For any number a, $-1 * a = a * -1 = \text{OPP}(a)$, or $-a$. The number a can be a negative number, so $\text{OPP}(a)$ or $-a$ can be a positive number. For example, if $a = -5$, then $-a = \text{OPP}(-5) = 5$.

Date ____________________ Time ____________________

LESSON 6•4 Fact Families for Multiplication and Division

A fact family is a group of four basic, related multiplication and division facts.

Example: The multiplication and division fact family for 6, 3, and 18 is made up of the following facts:

$6 * 3 = 18$ $18 / 6 = 3$
$3 * 6 = 18$ $18 / 3 = 6$

As you already know, when a positive number is divided by a positive number, the quotient is a positive number. Problems 1 and 2 will help you discover the rules for division with negative numbers. Complete the fact families. Check your answers with a calculator. Then complete each rule.

1. **a.** $5 * (-3) =$ ________

$-3 * 5 =$ ________

$-15 / (-3) =$ ________

$-15 / 5 =$ ________

b. $6 * (-8) =$ ________

$-8 * 6 =$ ________

$-48 / (-8) =$ ________

$-48 / 6 =$ ________

c. $5 * (-5) =$ ________

____

____

____

d. **Negative / Negative Rule:** When a negative number is divided by a negative number, the quotient is a ________ (positive or negative) number.

e. **Negative / Positive Rule:** When a negative number is divided by a positive number, the quotient is a ________ (positive or negative) number.

2. **a.** $-4 * (-3) =$ ________

$-3 * (-4) =$ ________

$12 / (-3) =$ ________

$12 / (-4) =$ ________

b. $-7 * (-5) =$ ________

____

____

____

c. $-2 * (-10) =$ ________

____

____

____

d. **Positive / Negative Rule:** When a positive number is divided by a negative number, the quotient is a ________ (positive or negative) number.

3. Solve. Check your answers with a calculator.

a. ________ $* (-4) = 24$ (*Think:* What number multiplied by -4 is equal to 24?)

b. ________ $* 9 = -81$

c. $-6 *$ ________ $= 48$

d. ________ $* (-3) = -27$

e. $-81 / 9 =$ ________

f. $48 / (-6) =$ ________

g. $-27 / (-3) =$ ________

Date Time

LESSON 6·4

*, / of Positive and Negative Numbers

A Multiplication Property

For all numbers *a* and *b,* if the values of *a* and *b* are both positive or both negative, then the product $a * b$ is a positive number. If one of the values is positive and the other is negative, then the product $a * b$ is a negative number.

A Division Property

For all numbers *a* and *b,* if the values of *a* and *b* are both positive or both negative, then the quotient a / b is a positive number. If one of the values is positive and the other is negative, then the quotient a / b is a negative number.

Solve. Use a calculator to check your answers.

1. $-7 * 8 =$ ________
2. $73 * (-45) =$ ________
3. ________ $\div (-10) = 70$
4. $\frac{1}{2} * (-\frac{3}{4}) =$ ________
5. $0.5 * (-15) =$ ________
6. ________ $* 3.3 = -3.3$
7. $-3 * 4 * (-7) =$ ________
8. ________ $* (-8) * (-3) = -48$
9. $-54 / 9 =$ ________
10. $36 / (-12) =$ ________
11. $-\frac{3}{5} \div (-\frac{4}{5}) =$ ________
12. $45 / (-5) / (-3) =$ ________
13. ________ $\div 15 = -6$
14. $72 / (-8) =$ ________
15. $-99 /$ ________ $= -11$
16. $\frac{1}{2} \div (-\frac{3}{4}) =$ ________
17. $-3 * (-4 + 6) =$ ________
18. $32 \div (-5 - 3) =$ ________
19. $(-9 * 4) + 6 =$ ________
20. $(-75 / 5) + (-20) =$ ________
21. $(-6 * 3) + (-6 * 5) =$ ________
22. $(4 * (-7)) - (4 * (-3)) =$ ________

Evaluate each expression for $y = -4$.

23. $3 - (-y) =$ ________
24. $-y / (-6) =$ ________
25. $y - (-7 + 3) =$ ________
26. $y - (y + 2) =$ ________
27. $(-8 * y) - 6 =$ ________
28. $(-8 * 6) - (-8 * y) =$ ________

Date Time

LESSON 6•4

Math Boxes

1. Write the reciprocal.

a. 5 ________

b. $\frac{2}{3}$ ________

c. $2\frac{4}{7}$ ________

d. 9.64 ________

2. Divide. Simplify if possible.

a. $\frac{3}{5} \div 1\frac{1}{4} =$ ________

b. $\frac{5}{6} \div 2\frac{1}{3} =$ ________

c. ________ $= 6\frac{4}{5} \div 2\frac{1}{2}$

d. ________ $= 20\frac{2}{5} \div 10\frac{1}{5}$

3. Lines *a* and *b* are parallel. Without using a protractor, find the degree measure of each numbered angle. Write each measure on the drawing.

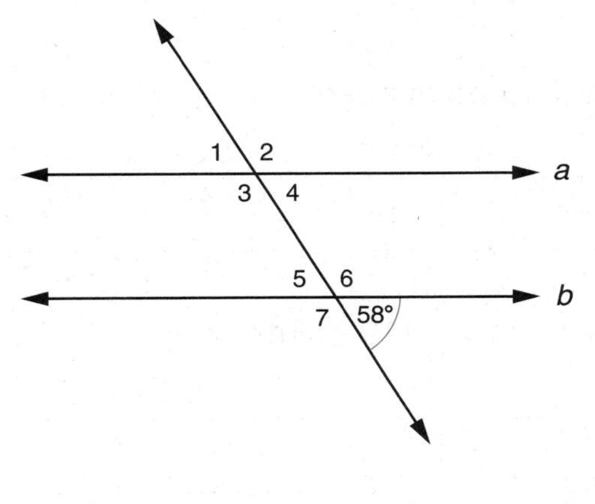

4. If 1 kilogram (kg) is about $2\frac{1}{5}$ pounds (lb), then 5 kg is about ________ lb.

Circle the best answer.

A. $\frac{11}{25}$

B. $7\frac{1}{5}$

C. 10

D. 11

5. Express each decimal as a percent.

a. 1.04 = ____________

b. ____________ = 0.0825

c. 0.0035 = ____________

d. ____________ = 4.0

6. Suppose you toss a penny and a nickel together. How many equally likely outcomes are there?

Complete the table.

Penny	Nickel
H	H

________ equally likely outcomes

Date ______________________ Time ______________________

LESSON 6•5 Scavenger Hunt

Use *Student Reference Book,* pages 2–24 and 94–106 to find answers to as many of these questions as you can. Try to get as high a score as possible.

1. How many rational numbers are there? (10 points) ______________________

2. Give an example of each of the following. (5 points each)

   a. A counting number ____________

   b. A negative rational number ____________

   c. A positive rational number ____________

   d. A real number ____________

   e. An integer ____________

   f. An irrational number ____________

3. Name two examples of uses of negative rational numbers. (5 points each)

   ____

   ____

4. Explain why numbers such as 4, $\frac{3}{5}$, and 3.5 are rational numbers. (10 points)

   ____

   ____

   ____

5. Explain why numbers such as π and $\sqrt{2}$ are irrational numbers. (10 points)

   ____

6. $n + n = n$ What is n? ____________ (15 points)

7. $k = \text{OPP}(k)$ What is k? ____________ (15 points)

8. $j * j = j$ Which two numbers could j be? ____________ (15 points each)

9. $a + (-a) =$ ____________ (15 points)

10. $b * \frac{1}{b} =$ ____________ (15 points)

Date Time

LESSON 6•5

Scavenger Hunt *continued*

11. Match each sentence in Column 1 with the property in Column 2 that it illustrates. (5 points each)

Column 1

A. $a + (b + c) = (a + b) + c$

B. $a + b = b + a$

C. $a * (b + c) = (a * b) + (a * c)$

D. $a * (b - c) = (a * b) - (a * c)$

E. $a * b = b * a$

F. $a * (b * c) = (a * b) * c$

Column 2

_______ Distributive Property of Multiplication over Subtraction

_______ Commutative Property of Addition

_______ Distributive Property of Multiplication over Addition

_______ Associative Property of Multiplication

_______ Commutative Property of Multiplication

_______ Associative Property of Addition

12. $-a > 0$. How can that be? (15 points)

____

____

13. Complete. (2 points each, except the last problem, which is worth 25 points)

OPP(1) = _______

OPP(OPP(1)) = _______

OPP(OPP(OPP(1))) = _______

OPP(OPP(OPP(OPP(1)))) = _______

OPP(OPP(OPP(OPP(OPP(OPP … (OPP(OPP(1)))))))) = _______

100 OPPs

Explain how you found the answer to the last problem. _______________________

____

____

Date ______________________ Time ______________________

LESSON 6•5 Scavenger Hunt *continued*

14. Is 5^{-2} a positive or negative number? Explain. (15 points)

____

____

____

15. Two numbers are their own reciprocals. What are they? ______________ (15 points each)

16. What number has no reciprocal? __________ (15 points)

Number Stories

1. Diana wants to make a 15 ft by 20 ft section of her yard into a garden. She will plant flowers in $\frac{2}{3}$ of the garden and vegetables in the rest of the garden. How many square feet of vegetable garden will she have?

____

Explain how you got your answer. ____

____

____

____

____

2. Leo is in charge of buying hot dogs for his school's family math night. Out of 300 parents and children, he expects about $\frac{3}{5}$ of them to attend. Hot dogs are sold 8 in a package, and Leo figures he will need to buy 22 packages so that each person can have 1 hot dog.

a. How do you think he calculated to get 22 packages?

____

____

____

b. Will Leo have enough hot dogs? ____

____

Date ______________________ Time ______________________

LESSON 6•5 Math Boxes

1. Simplify.

a. $\frac{3}{4}$ of 80 ________

b. $\frac{9}{8}$ of 2 ________

c. $\frac{2}{3}$ of $3\frac{1}{2}$ ________

d. $\frac{3}{8}$ of $\frac{4}{9}$ ________

2. Divide. Simplify if possible.

a. $\frac{8}{9} \div \frac{3}{4} =$ ________

b. $\frac{7}{8} \div \frac{1}{3} =$ ________

c. $\frac{6}{9} \div \frac{1}{2} =$ ________

d. $\frac{8}{24} \div \frac{4}{24} =$ ________

3. Give a ballpark estimate for each quotient.

a. $643.27 \div 5$ ________

b. $\frac{728.09}{7}$ ________

c. $432.67 \div 82$ ________

d. 2,091.05 / 53 ________

4. Complete each sentence using an algebraic expression.

a. If Mark earns *t* dollars for each hour he tutors, then he earns

____________ dollars when he tutors

for $3\frac{1}{2}$ hours.

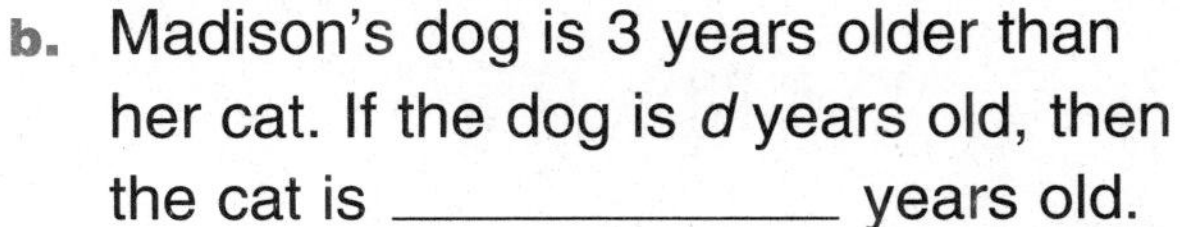

b. Madison's dog is 3 years older than her cat. If the dog is *d* years old, then the cat is ____________ years old.

5. Write each decimal as a fraction in simplest form.

a. 0.06 = ________

b. 0.52 = ________

c. 0.09 = ________

d. 0.64 = ________

6. Roll two 6-sided dice, one red, one green. Give the probability of rolling the following totals.

a. 11 ________ **b.** 7 ________

c. 0 ________ **d.** 3 or 4 ________

e. an even number ________

Date Time

LESSON 6•6

Order of Operations

Evaluate each expression. Show your work. Then compare your results to those of your partner.

1. $4 * 6 + 3 =$ __________

2. $33 - 16 / 4 =$ __________

3. $4 * 7 - (3 + 5) =$ __________

4. $24 / 6 * 4 =$ __________

5. $8 * 4 + 49 \div 7 =$ __________

6. $9 * 6 \div 3 + 28 =$ __________

7. $7 - 5 + 13 - 23 - 17 =$ __________

8. $100 - 50 \div 2 + 4 * 5 =$ __________

9. $7 / 7 * 4 + 3^2 =$ __________

10. $12 * 2^2 - 3^3 =$ __________

Date Time

LESSON 6•6

Order of Operations *continued*

11. $10^{-1} + 16 - 0.5 * 12 =$ ________

12. $((\frac{1}{2} \div \frac{1}{4}) + 3) * 6 - 3^3 =$ ________

13. $-(-8) - (-4) * 6 - (-12) / 4 =$ ________

14. $-4 + (-18) / 6 + (-3 * -3 - 5) =$ ________

Try This

15. $-5(-6 - (-3)) / 7.5 =$ ________

16. $-(\frac{3}{4} \div \frac{1}{2}) + \frac{1}{2} - (\frac{1}{2} * (-\frac{1}{2})) =$ ________

17. Evaluate the following expressions for $x = -2$.

a. $x * -x + 14 / 2 =$ ________

b. $-x * (6 + x) - 3^3 / 9 =$ ________

Date Time

LESSON 6•6

Math Boxes

1. Solve. Simplify your answers.

a. ________ $= 8 \div 10\frac{2}{3}$

b. $4\frac{1}{2} \div 1\frac{5}{7} =$ ________

c. ________ $= 7\frac{3}{10} \div 5$

2. Multiply or divide.

a. $-10(-14.35) =$ ________

b. $4 * 3 * (-5) =$ ________

c. ________ $= \frac{280}{-4}$

3. Triangles *JKL* and *MNO* are congruent.

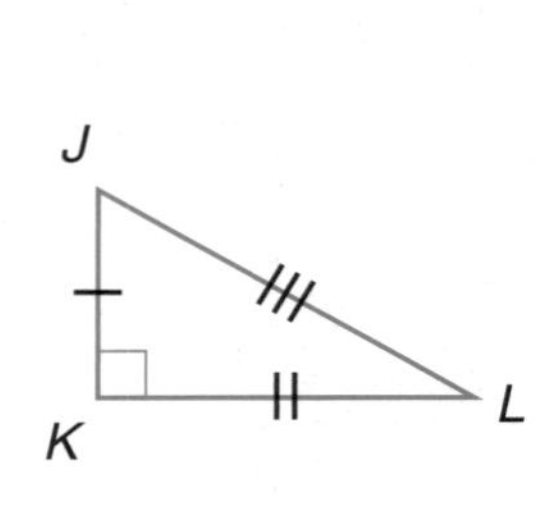

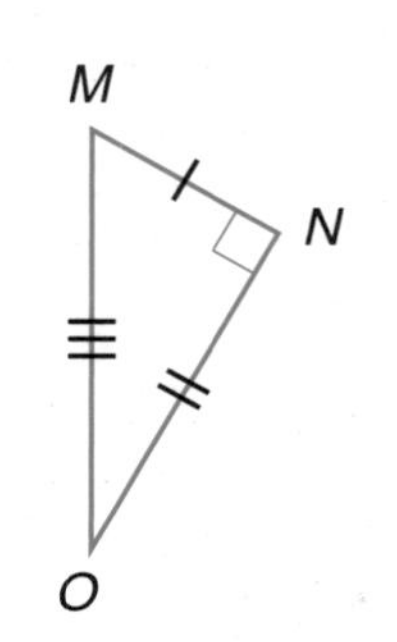

Which side corresponds with $\overline{JL}$?

____

4. Label the axes of this mystery graph and describe a situation it might represent.

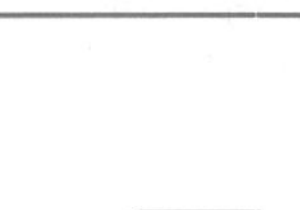

x-axis ________________________

y-axis ________________________

Situation ________________________

____

____

5. Two dice are tossed. Some possible outcomes appear in the table below.

Complete the table.

(1,1)	(1,2)	(1,3)	(1,4)	(1,5)	(1,6)
			(3,4)		
(4,1)					
				(6,5)	

a. How many equally likely outcomes are there? ________

b. What is the probability of tossing a multiple of 2 on both dice? ________

c. What is the probability of tossing a composite number on the first die and a prime number on the second die? ________

Date Time

LESSON 6•7 Math Boxes

1. Simplify.

a. $\frac{4}{7}$ of 84 ________

b. $\frac{1}{20}$ of 35 ________

c. $\frac{8}{3}$ of $9\frac{3}{4}$ ________

d. $\frac{1}{3}$ of $\frac{51}{68}$ ________

2. Divide. Simplify if possible.

a. $\frac{7}{12} \div \frac{2}{5} =$ ________

b. $2\frac{5}{8} \div 3 =$ ________

c. $3 \div 6\frac{1}{4} =$ ________

d. $4\frac{1}{2} \div 2\frac{7}{10} =$ ________

3. Give a ballpark estimate for each quotient.

a. 137.8 ÷ 15 ________

b. 248.19 / 12 ________

c. 4,507.08 ÷ 89.76 ________

d. 0.6 / 14.7 ________

SRB 261

4. Complete each sentence using an algebraic expression.

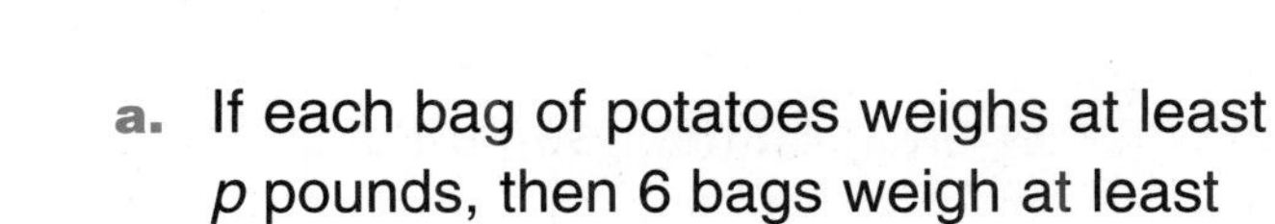

a. If each bag of potatoes weighs at least p pounds, then 6 bags weigh at least

__________ pounds.

b. Jack is 6 inches taller than Michael. If Jack is h inches tall, then Michael is

__________ inches tall.

5. Which fraction is equivalent to 2.015? Choose the best answer.

○ $\frac{2,015}{10,000}$

○ $\frac{4,030}{20,000}$

○ $\frac{403}{200}$

○ $\frac{2,015}{100}$

6. You draw one card at random from a regular deck of 52 playing cards (no jokers). What is the chance of drawing:

a. a 4? ________

b. a card with a prime number? ________

c. a face card (jack, queen, or king)? ________

d. an even-numbered black card? ________

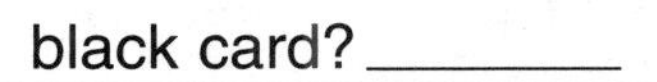

Date Time

LESSON 6·7 Number Sentences

Translate the word sentences below into number sentences. Study the first one.

1. Three times five is equal to fifteen. $3 * 5 = 15$
2. Nine increased by seven is less than twenty-nine. ______________
3. Thirteen is not equal to nine more than twenty. ______________
4. The product of eight and six is less than or equal to the sum of twenty and thirty.

   ____
5. Thirty-seven increased by twelve is greater than fifty decreased by ten.

   ____
6. Nineteen is less than or equal to nineteen. ______________

Tell whether each number sentence is true or false.

7. $3 * 21 = 63$ ______
8. $(3 * 4) + 7 = 19$ ______
9. $42 - 12 / 6 > 5$ ______
10. $8 \geq 7 + 1$ ______
11. $24 / 4 + 2 = 8$ ______
12. $9 / (8 - 5) \leq 3$ ______
13. $21 > (7 * 3) + 5$ ______
14. $8 * 7 \leq 72$ ______
15. $63 / 7 \neq 8$ ______
16. $35 + 5 * 8 = 320$ ______

Insert parentheses so each number sentence is true.

17. $5 * 8 + 4 - 2 = 42$
18. $7 * 9 - 6 = 21$
19. $10 + 2 * 6 < 24$
20. $9 - 7 / 7 = 8$
21. $33 - 24 / 3 \geq 25$
22. $36 / 7 + 2 * 3 = 12$
23. $3 * 4 + 3 > 5 * 3 + 3$
24. $48 / 8 + 4 \neq 100 / 10$

LESSON 6•7

Number Sentences *continued*

25. Write three true and three false number sentences. Trade journals with your partner and determine which sentences are true and which are false.

Number Sentence	True or false?

Try This

26. The word HOPE is printed in shaded block letters inside a 15 ft by 5 ft rectangular billboard. What is the area of the unshaded portion of the billboard?

____

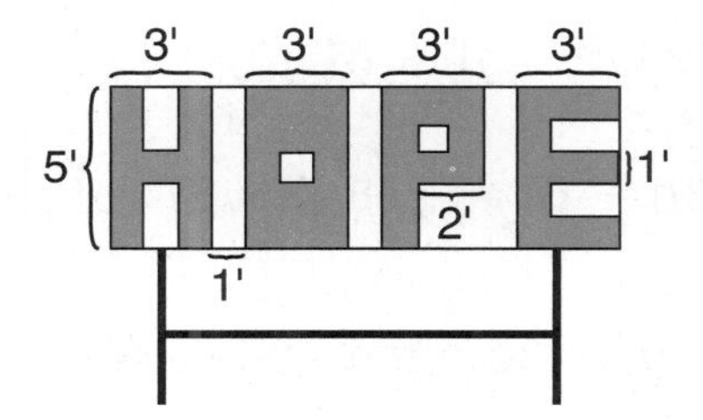

27. Square corners, 6 centimeters on a side, are removed from a 36 cm by 42 cm piece of cardboard. The cardboard is then folded to form an open box. What is the surface area of the inside of the box?

____

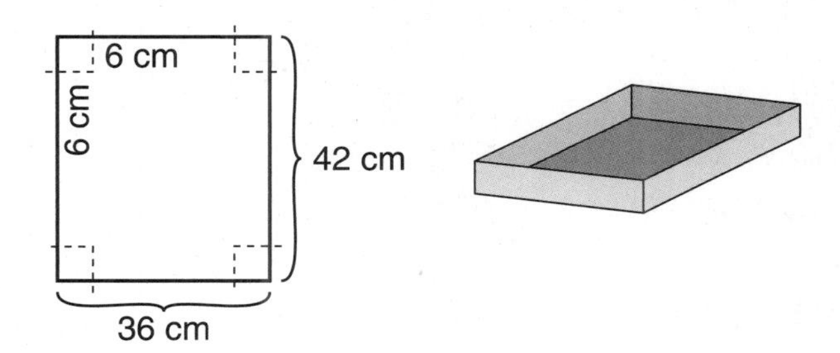

28. Pennies tossed onto the gameboard at the right have an equal chance of landing anywhere on the board. If 60% of the pennies land inside the smaller square, what is the length of a side *s* of the smaller square to the nearest inch?

____

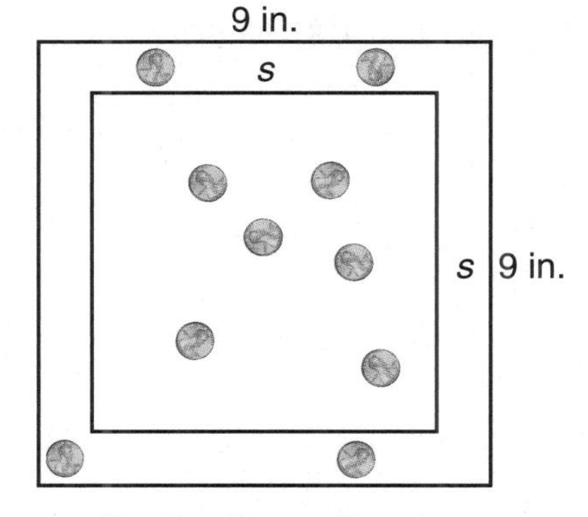

Try the Penny Toss!

Date Time

LESSON 6•8 Solving Equations

Find the solution to each equation. Write a number sentence with the solution in place of the variable. Check that the number sentence is true.

	Equation	Solution	Number Sentence
1.	$12 + x = 32$	______	______
2.	$y + 89 = 93$	______	______
3.	$b - 32 = 15$	______	______
4.	$m * 8 = 35 - 19$	______	______
5.	$p + (4 * 9) = 55$	______	______
6.	$42 = 7 * (a - 4)$	______	______
7.	$(9 + w) / 2 = 6 + (6 / 6)$	______	______
8.	$4 + (3n - 6) = 1 + (3 * 6)$	______	______

Find the solution to each equation.

9. $4 * 6 = 35 - t$ ______

10. $9 * (11 - c) = 81$ ______

11. $17 - 11 = k / 8$ ______

12. $(m + 14) / 4 = 6$ ______

13. $36 / 9 = 2 + p$ ______

14. $23 - a = 15$ ______

15. $(3 * p) + 5 = 26$ ______

16. $2 - d = 3 * 4$ ______

17. Make up four equations whose solutions are whole numbers. Ask your partner to solve each one.

	Equation	Solution
a.	______	______
b.	______	______
c.	______	______
d.	______	______

Date Time

LESSON 6•8 Math Boxes

1. Solve. Simplify your answers.

a. ________ $= -5\frac{1}{2} \div \frac{3}{4}$

b. $6 \div (-1\frac{5}{16}) =$ ________

c. ________ $= -3\frac{1}{8} \div (-4\frac{1}{8})$

2. Multiply or divide.

a. $(-3)^3 =$ ________

b. $0.4(-0.5) =$ ________

c. ________ $= \frac{-352}{-11}$

3. Triangles *DAB* and *BCD* are congruent.

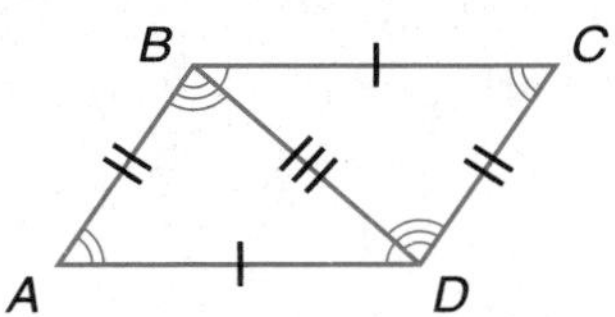

Which is a pair of corresponding angles? Circle the best answer.

A. ∠*ABD* and ∠*ABC*

B. ∠*BAD* and ∠*DCB*

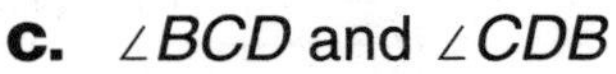

C. ∠*BCD* and ∠*CDB*

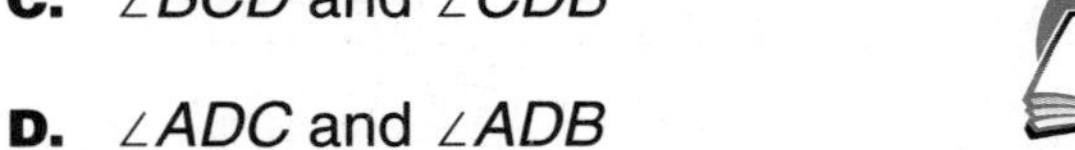

D. ∠*ADC* and ∠*ADB*

4. Label the axes of this mystery graph and describe a situation it might represent.

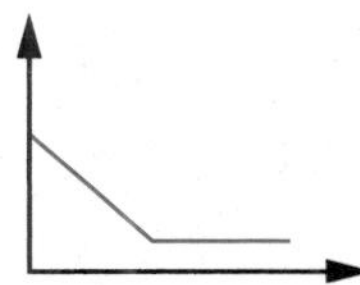

x-axis ________________________

y-axis ________________________

Situation ________________________

____

5. You spin the spinner shown at the right.

a. How many equally likely outcomes are there? ________

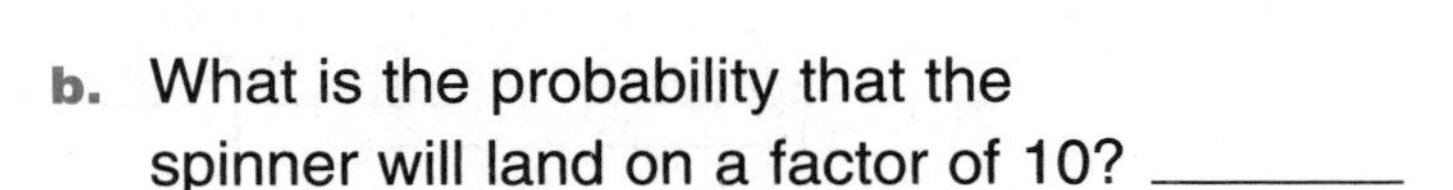

b. What is the probability that the spinner will land on a factor of 10? ________

c. What is the probability that the spinner will land on a multiple of 3 or a multiple of 2? ________

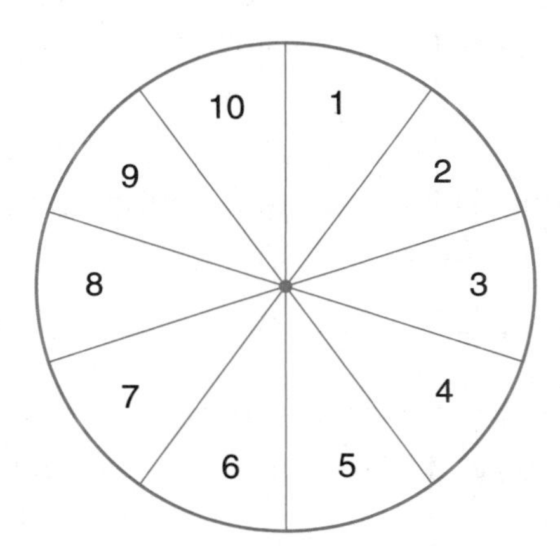

LESSON 6•9 Pan-Balance Problems

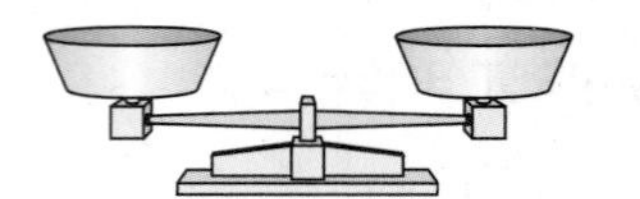

A pan balance can be used to compare the weights of objects or to weigh objects. If the objects in one pan weigh as much as those in the other pan, the pans will balance.

The diagram at the right shows a balanced pan balance.

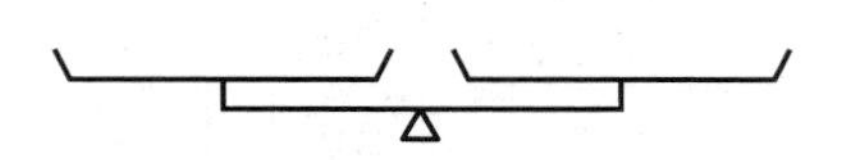

Example: In each of the diagrams below, the pans are balanced. Your job is to figure out how many marbles weigh as much as an orange. When moving the oranges and marbles, follow these simple rules:

- Whatever you do, the pans must always remain balanced.
- You must do the same thing to both pans.

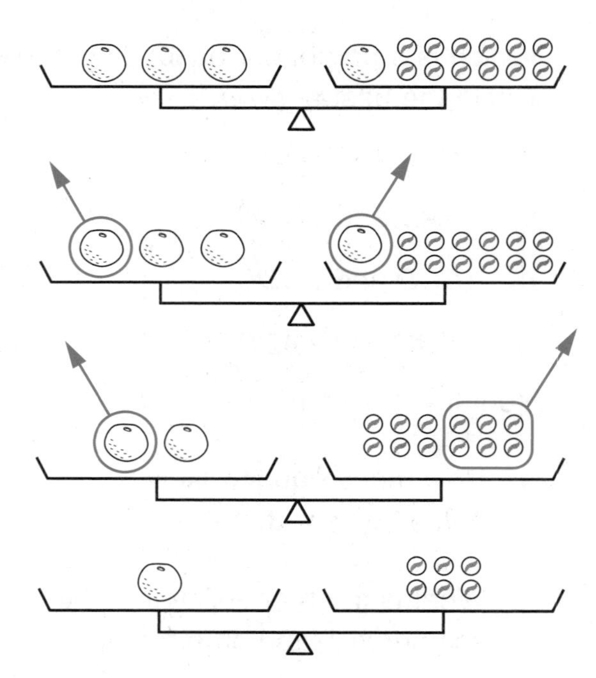

The pan balance shows that 3 oranges weigh as much as 1 orange and 12 marbles.

If you remove 1 orange from each pan, the pans remain balanced.

If you then remove half of the objects from each pan, the pans will still be balanced.

Success! One orange weighs as much as 6 marbles.

Solve the pan-balance problems with a partner. Be ready to share your strategies with the class.

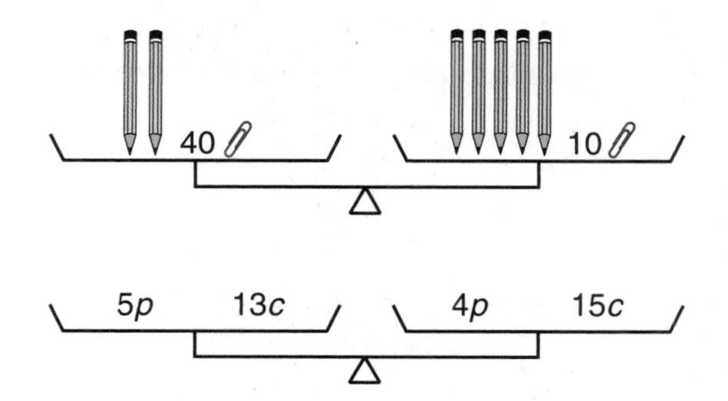

1. One pencil weighs as much as ________ paper clips.

2. One p (pencil) weighs as much as ________ cs (paper clips).

Date Time

LESSON 6•9

Pan-Balance Problems *continued*

Solve these pan-balance problems. In each figure, the two pans are balanced.

3. One banana weighs as much as ________ marbles.

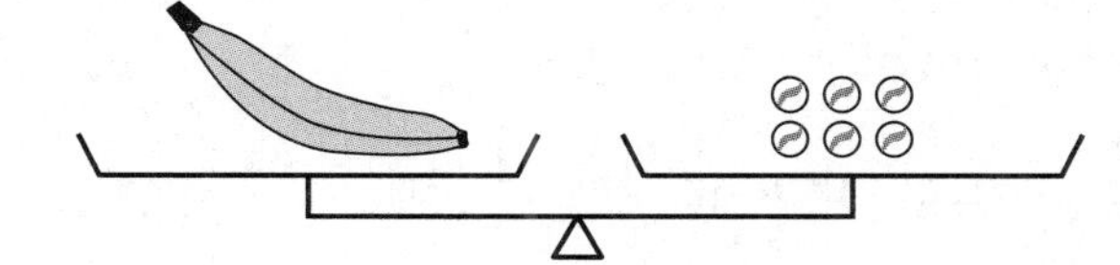

4. One cube weighs as much as ________ paper clips.

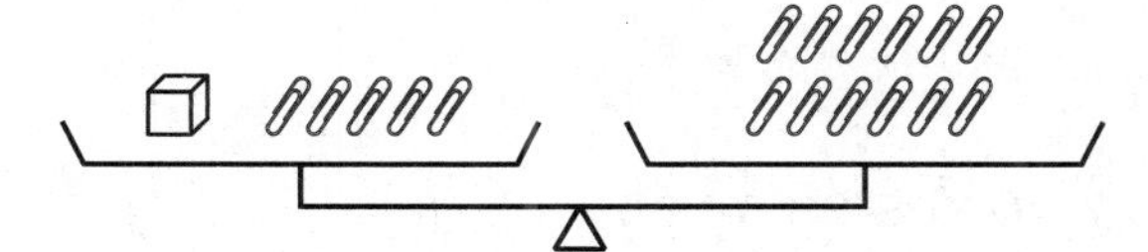

5. One cube weighs as much as ________ marbles.

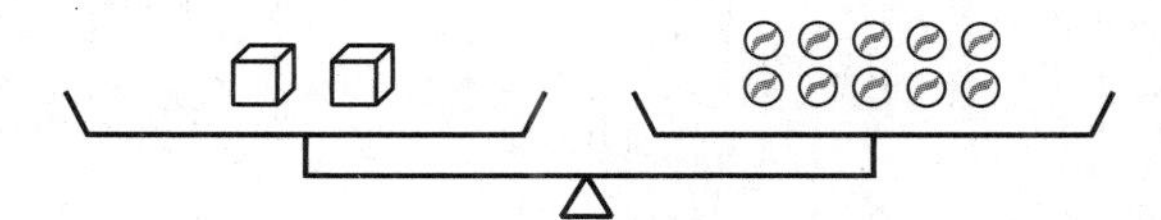

6. One triangle weighs as much as ________ squares.

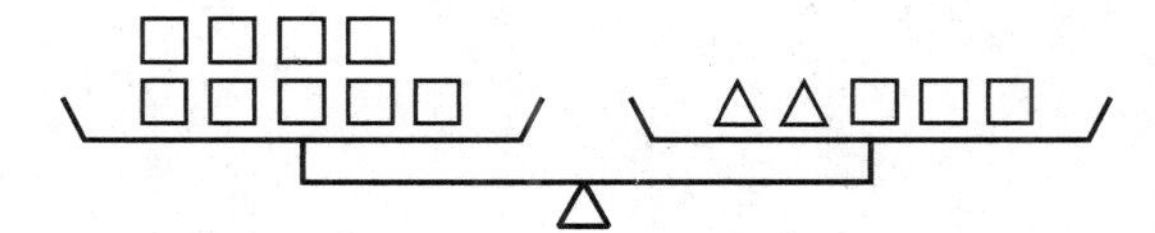

7. One orange weighs as much as ________ paper clips.

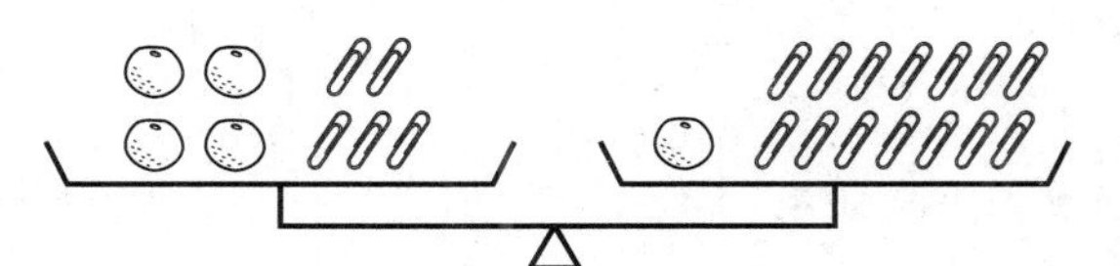

8. One and one-half cantaloupes weigh as much as ________ apples.

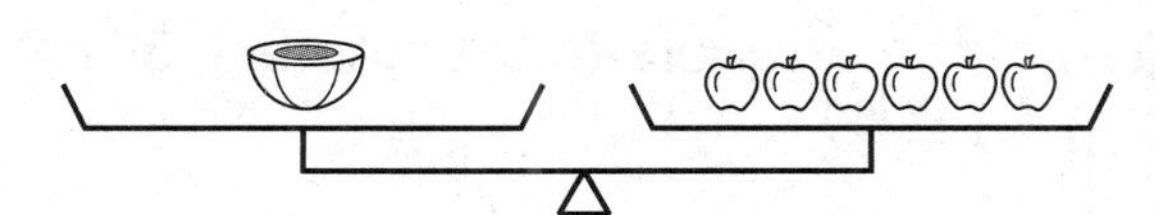

LESSON 6•9

Pan-Balance Problems *continued*

Reminder: 4□ or 4 * □ are other ways to write □ + □ + □ + □.

9. One cube weighs as much

as ________ coins.

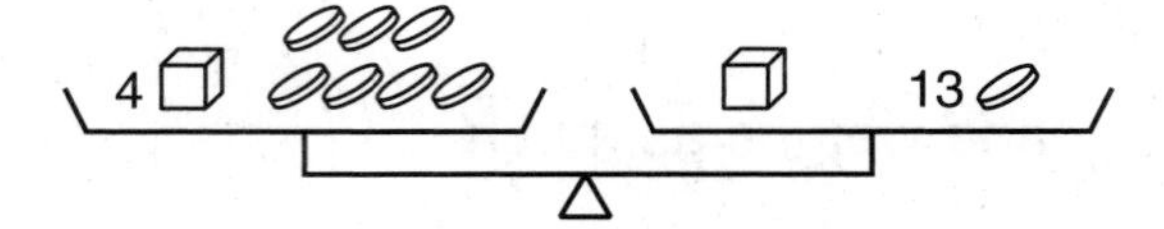

10. One *p* weighs as much

as ________ balls.

11. One *b* weighs as much

as ________ *k*s.

Check your answers.

- The sum of the answers to Problems 3 and 6 is equal to the square root of 81.
- The product of the answers to Problems 8 and 9 is 36.
- The sum of the answers to Problems 4, 5, and 11 is the solution to the equation $4n = 2^6$.
- The product of the answers to Problems 7, 9, and 11 is 24.

Date Time

LESSON 6•9

Math Boxes

1. Use the order of operations to evaluate each expression.

a. $\frac{1}{2} * (-26) - 3^2 =$ ________

b. $-3 - 14 \div 7 =$ ________

c. ________ $= (-8) * (-8) - (-8)$

d. $-5 - (-30) \div 3 =$ ________

2. Tell whether each statement is true or false.

a. If a $< b$, then $a - 3 < b - 3$. ________

b. If $m < n$, then $m + 8 > n * 8$. ________

c. If $x > y$, then $x * 0 > y * 0$. ________

3. You can use a formula to calculate about how long it will take a falling object to reach the bottom of a well.

Formula: $t = \frac{1}{4} * \sqrt{d}$

(This formula does not account for air resistance.)

- d is the distance in feet the object falls.
- t is the time in seconds it takes the object to reach the bottom.

About how long would it take a bowling ball to hit the bottom of a well 100 ft deep?

________ seconds

4. Lines a and b are parallel.

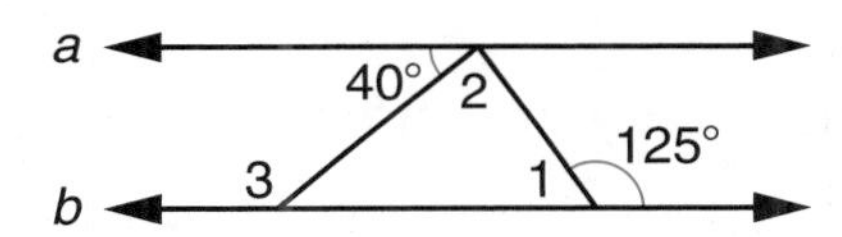

m∠1 = ________

m∠2 = ________

m∠3 = ________

5. Solve mentally.

a. 10% of 82 = ________

b. 5% of 44 = ________

c. 15% of 90 = ________

6. Suppose you toss a coin 10 times and get 10 HEADS. What is the probability of getting HEADS on the 11th toss?

Probability = ________

Date ______________________ Time ______________________

LESSON 6•10

Pan-Balance Equations

1. Start with the original pan-balance equation. Do the first operation on both sides of the pan balance and write the result on the second pan balance. Do the second operation on both sides of the second pan balance and write the result on the third pan balance. Complete the fourth pan balance in the same way.

Original pan-balance equation $x = 10$

Operation

(in words)	(abbreviation)	
Multiply by 3.	M 3	______ = ______
Add 18.	A 18	______ = ______
Add $2x$.	A $2x$	______ = ______

Equations that have the same solution are called **equivalent equations.**

2. Check that the pan-balance equations above are equivalent equations by making sure that 10 is the solution to each equation.

3. Now do the opposite of what you did in Problem 1. Record the operation used to obtain the results on each pan balance.

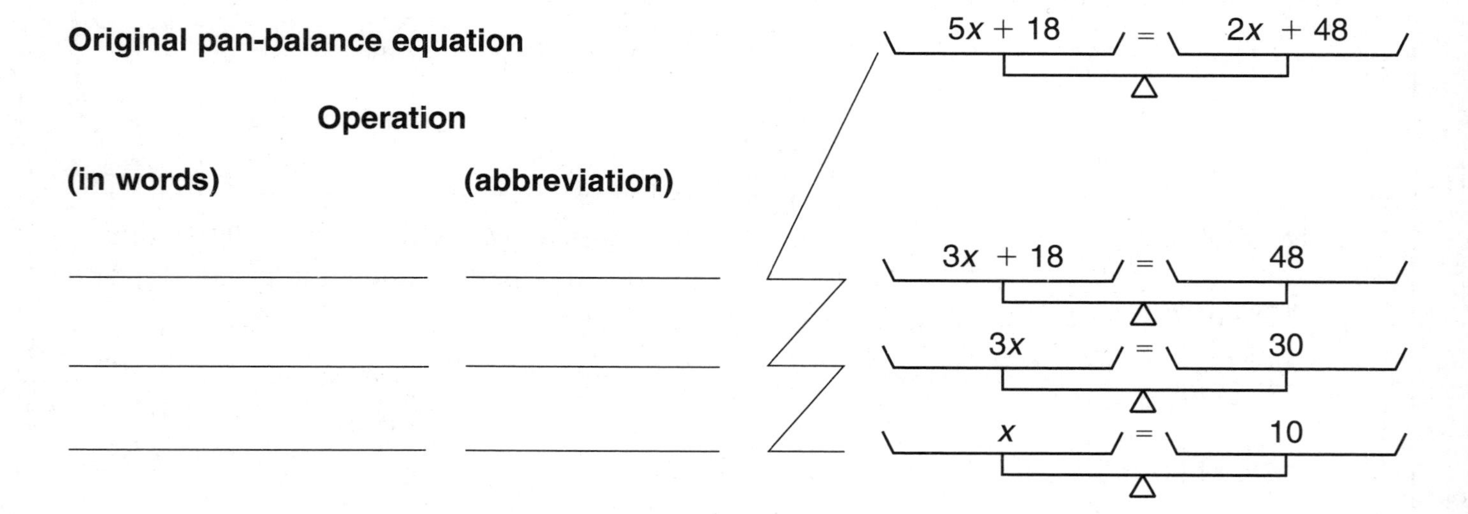

Date Time

LESSON 6•10 Pan-Balance Equations *continued*

4. Record the results of the operation on each pan, as in Problem 1.

SRB 250–252

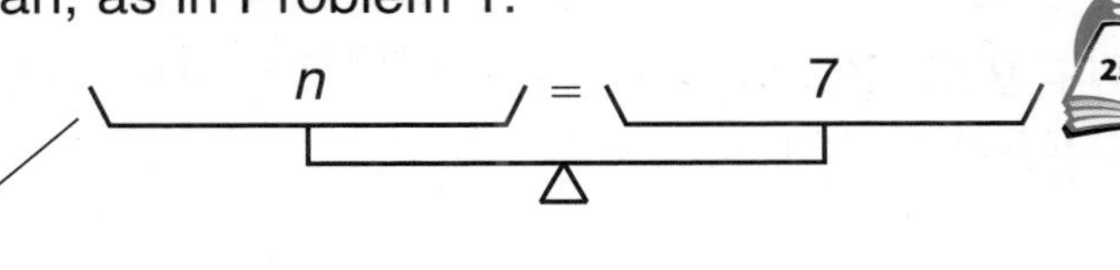

Original pan-balance equation

Operation	
(in words)	**(abbreviation)**
Subtract 2.	S 2
Multiply by 4.	M 4
Add $2n$.	A $2n$

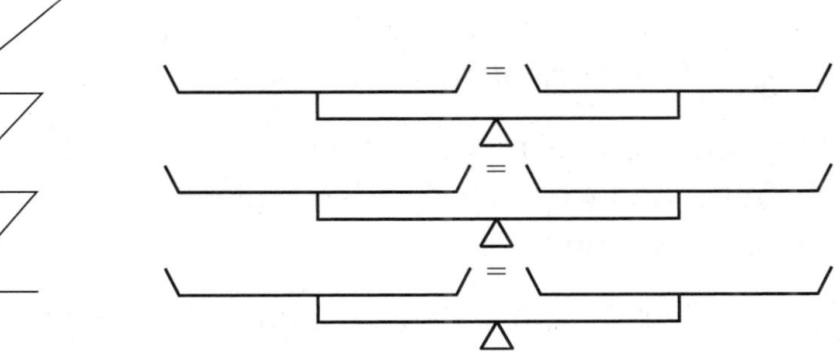

5. Check that 7 is the solution to each pan-balance equation in Problem 4.

In Problems 6 and 7, record the operation that was used to obtain the results on each pan balance, as you did in Problem 3.

6. **Original pan-balance equation**

Operation

(in words) **(abbreviation)**

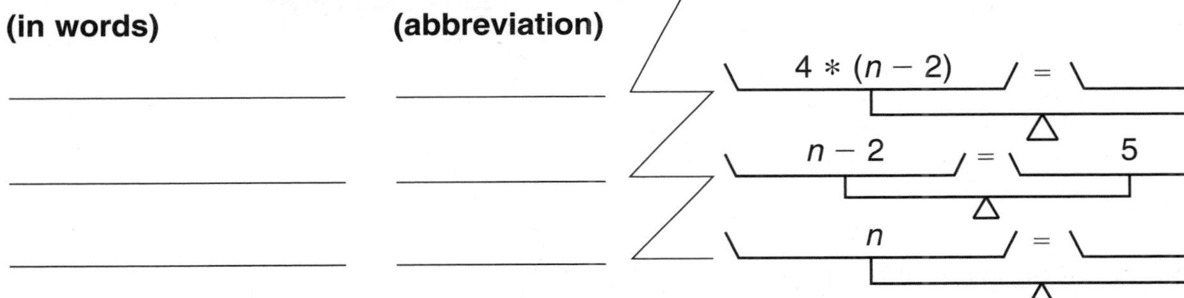

7. **Original pan-balance equation**

Operation

(in words) **(abbreviation)**

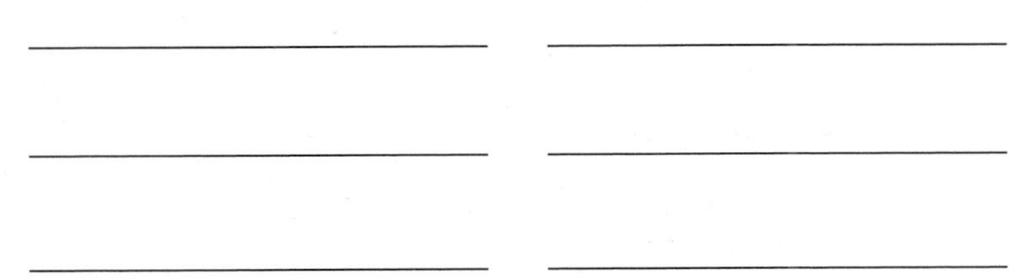

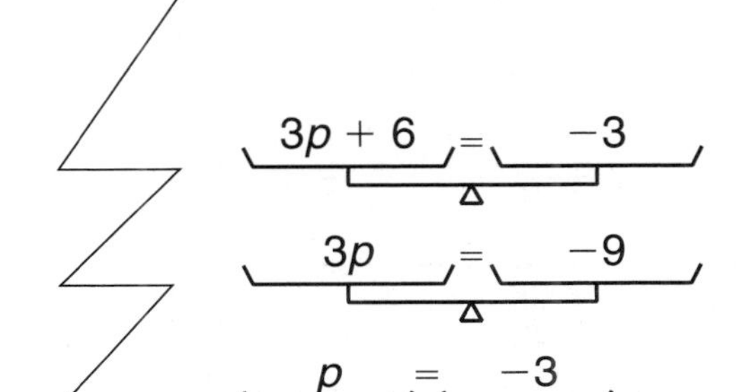

8. Check that 7 is the solution to each pan-balance equation in Problem 6 and that -3 is the solution to each pan-balance equation in Problem 7.

Date Time

LESSON 6•10

Inventing and Solving Equations

Work in groups of three. Each of you will invent two equations and then ask the other two group members to solve them. Show your solutions on page 237. Here is what to do for each equation.

Step 1 Choose any positive or negative integer and record it on the first line to complete the original equation.

Step 2 Apply any operation you wish to both sides of the equation. Record the operation and write the new (equivalent) equation on the lines below the original equation.

Step 3 Repeat Step 2. Apply a new operation and show the new equation that results.

Step 4 Check your work by substituting the original value of x in each equation you have written. You should get a true number sentence every time.

Step 5 Give the other members of your group the final equation to solve.

1. Make up an equation from two equivalent equations.

Original equation x = ____________
Selected integer

Operation

____________ ____________ = ____________

____________ ____________ = ____________

2. Make up an equation from three equivalent equations.

Original equation x = ____________
Selected integer

Operation

____________ ____________ = ____________

____________ ____________ = ____________

____________ ____________ = ____________

Date Time

Inventing and Solving Equations *continued*

Use this page to solve your group members' equations.

First, record the equation. Then solve it. For each step, record the operation you use and the equation that results. Check your solution by substituting it for the variable in the original equation. Finally, compare the steps you used to solve the group member's equation to the steps he or she used in inventing the equation.

1. Member's equation ______ = ______

Operation

______ ______ = ______

______ ______ = ______

______ ______ = ______

2. Member's equation ______ = ______

Operation

______ ______ = ______

______ ______ = ______

______ ______ = ______

3. Member's equation ______ = ______

Operation

______ ______ = ______

______ ______ = ______

______ ______ = ______

4. Member's equation ______ = ______

Operation

______ ______ = ______

______ ______ = ______

______ ______ = ______

Date Time

LESSON 6•10 Math Boxes

1. Find the solution to each equation.

a. $-9 + d = 15$ $d =$ ______

b. $52 = -48 + m$ $m =$ ______

c. $4 * (12 - x) = -4$ $x =$ ______

d. $p + (-9) = -5$ $p =$ ______

2. Solve the pan-balance problem.

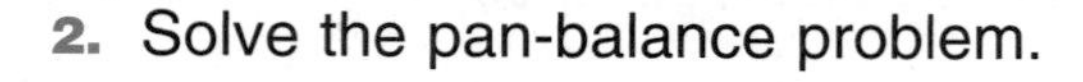
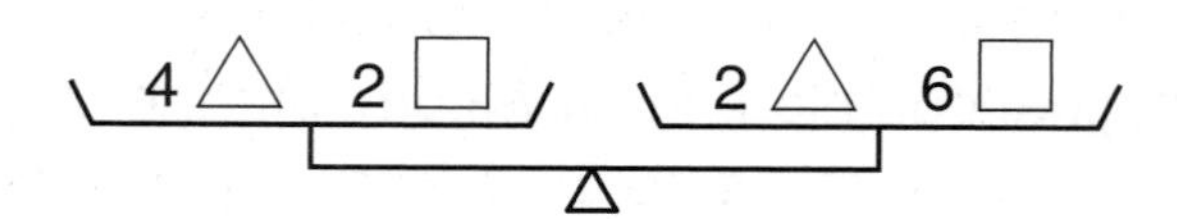

One △ weighs as much as ______ □s.

3. Complete the table. Then graph the data and connect the points.

Heather earns $0.35 for each paper flower she makes for the school fun fair.

Rule:
Earnings = $0.35 * number of flowers

Flowers (*f*)	Earnings ($) (0.35 * *f*)
1	
2	
	1.05
5	
	2.10

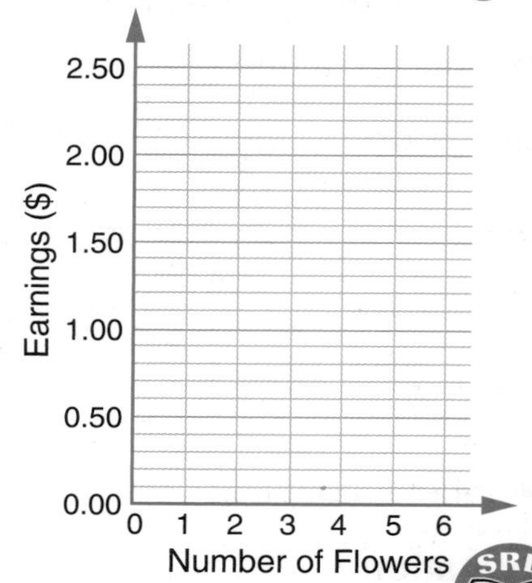

4. Suppose you roll a regular 6-sided die 90 times. About how many times would you expect to roll a 3?

About ______ times

5. How many arrangements of the letters *B, O,* and *X* are possible if you use each letter only once in each arrangement? List the arrangements.

______ possible arrangements

______

______

______

SRB 156

Date Time

LESSON 6•11

Math Boxes

1. Use the order of operations to evaluate each expression.

a. $(-3 * 7) + (-2 * 4) =$ ________

b. $(-4)^3 \div 4 - (-2) =$ ________

c. ________ $= \frac{1}{3}(9 * -3) + 18$

d. $\frac{14-40}{-2} =$ ________

2. Which of the following statements is *not* true? Circle the best answer.

A The order in which two numbers are multiplied does not change the product.

B The product of any number and -1 is a negative number.

C The product of any number and -1 is the opposite of the number.

SRB 104 105

3. Suppose N is a 2-digit whole number that ends in 5, such as 15. It is easy to find the square of N by using the formula

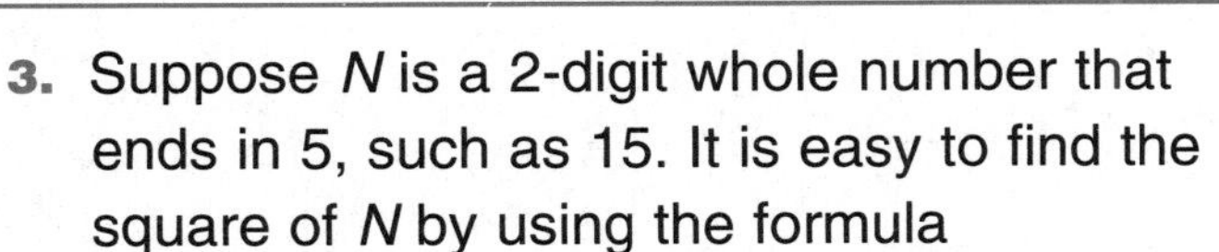

$$N^2 = (t * (t + 1) * 100) + 25$$

where t is the tens digit of the number you are squaring. Use this formula to find the following. (*Hint:* In Problem a, $t = 3$.)

a. $35^2 =$ ____________

b. $75^2 =$ ____________

c. $95^2 =$ ____________

4. Lines l and m are parallel.
Lines p and q are parallel.

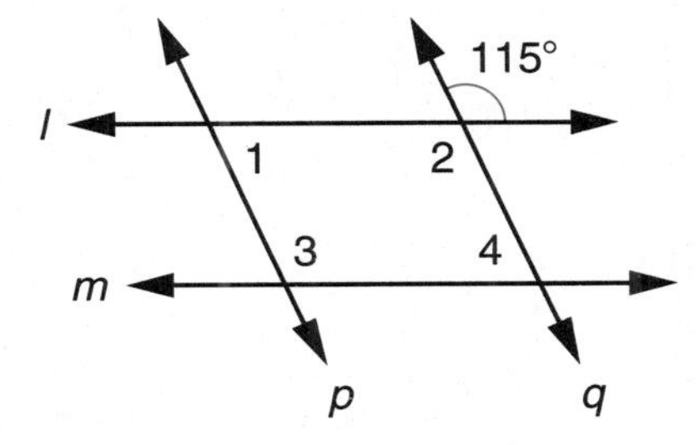

$m\angle 1 + m\angle 4 =$ ________

5. Solve mentally.

a. 59% of 100 = ______

b. 25% of 30 = ______

c. 75% of 88 = ______

6. A bag contains 4 black marbles, 2 white marbles, and 1 gray marble. If a marble is drawn at random, what is the probability of:

a. drawing a black marble? ______

b. drawing a white marble? ______

c. drawing a gray or a white marble?

$\frac{\quad}{\quad} + \frac{\quad}{\quad} = \frac{\quad}{\quad}$

SRB 153

Date Time

LESSON 6•11

Solving Equations

Solve the following equations.

SRB 250–252

Example:

$3x + 5 = 14$

Original equation $3x + 5 = 14$

Operation

S 5 $3x = 9$

D 3 $x = 3$

Check $(3 * 3) + 5 = 14$; true

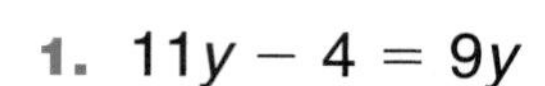

1. $11y - 4 = 9y$

Original equation

____

Operation

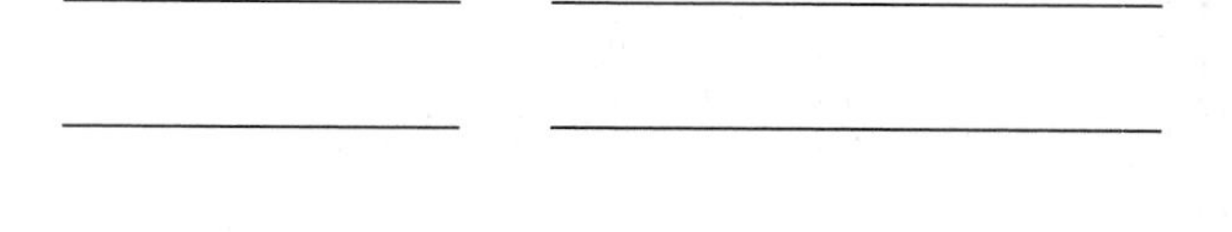

______ __________

______ __________

______ __________

Check ________________

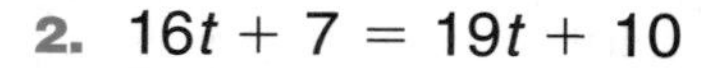

2. $16t + 7 = 19t + 10$

Original equation

____

Operation

______ __________

______ __________

______ __________

Check ________________

3. $12n - 5 = 9n - 2$

Original equation

____

Operation

______ __________

______ __________

______ __________

Check ________________

4. $8k - 6 = 10k + 6$

Original equation

____

Operation

______ __________

______ __________

______ __________

Check ________________

5. $3b + 7.1 = 2.5b + 11.5$

Original equation

____

Operation

______ __________

______ __________

______ __________

Check ________________

Date Time

LESSON 6•11

Solving Equations *continued*

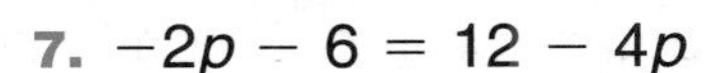

6. $8 - 3h = 5h + 1$

Original equation

____

Operation

__________ ______________

__________ ______________

__________ ______________

Check ______________________

7. $-2p - 6 = 12 - 4p$

Original equation

____

Operation

__________ ______________

__________ ______________

__________ ______________

Check ______________________

8. $\frac{1}{4}r + 9 = 10 - \frac{3}{4}r$

Original equation

____

Operation

__________ ______________

__________ ______________

Check ______________________

9. $\frac{2}{3}u - 7 = 9 - \frac{2}{3}u$

Original equation

____

Operation

__________ ______________

__________ ______________

__________ ______________

Check ______________________

Try This

Two equations are equivalent if they have the same solution. Circle each pair of equivalent equations. Write the solution to the equations if they are equivalent.

10. $z = 5$

$3z - 8 = 2z - 3$

Solution ______________

11. $d + 5 = 8$

$6 - 2d = 9 - 3d$

Solution ______________

12. $v + 1 = 2v + 2$

$3v - 8 = 2v - 3$

Solution ______________

13. $t = 4$

$(5t + 3) - 2(t + 3) = 29 - 5t$

Solution ______________

Date Time

LESSON 6•12

Dividing Decimals by Decimals: Part 1

When you multiply the numerator and denominator of a fraction by the same nonzero number, you rename the fraction without changing its value.
For example: $\frac{3}{5} * \frac{10}{10} = \frac{30}{50}$, so $\frac{3}{5} = \frac{30}{50}$

In general, you can think of a fraction as a division problem. The fraction $\frac{3}{5}$ equals $3 \div 5$, or 5)3. The numerator of the fraction is the dividend; the denominator is the divisor. As you do with a fraction, you can multiply the dividend and divisor by the same number without changing the value of the quotient.

Study the patterns in the table below.

Fraction	Division Problem	Quotient
$\frac{6.08}{0.08}$	0.08)6.08	76
$\frac{6.08}{0.08} * \frac{10}{10}$	0.8)60.8	76
$\frac{6.08}{0.08} * \frac{100}{100}$	8)608	76

What do you notice about the quotient when you multiply the dividend and the divisor by the same number?

____

Rename each division problem so the divisor is a whole number. Then solve the equivalent problem using partial-quotients division or another method.

Example:

Equivalent Problem

0.005)0.015 = 5)15   Quotient = 3

1. 0.004)2.05 = ____________________

Equivalent Problem

Quotient = ____________

2. 0.3)7.08 = ____________________

Equivalent Problem

Quotient = ____________

Date Time

LESSON 6•12

Dividing Decimals by Decimals *continued*

Rename each division problem so the divisor is a whole number. Then solve the equivalent problem using partial-quotients division or another method.

Equivalent Problem

3. $0.14\overline{)294}$ = ______________

Quotient = __________

Equivalent Problem

4. $0.013\overline{)6.24}$ = ______________

Quotient = __________

Equivalent Problem

5. $0.46\overline{)33.58}$ = ______________

Quotient = __________

Equivalent Problem

6. $1.67\overline{)13.36}$ = ______________

Quotient = __________

Date Time

Inequalities

1. Name two solutions of each inequality.

SRB 244

a. $15 > r$ ____________ **b.** $8 < m$ ____________

c. $t \geq 56$ ____________ **d.** $15 - 11 \leq p$ ____________

e. $\frac{21}{7} > y$ ____________ **f.** $w > -3$ ____________

g. $6.5 > 3 * d$ ____________ **h.** $g < 0.5$ ____________

2. Name two numbers that are *not* solutions of each inequality.

a. $(7 + 3) * q > 40$ ____________ **b.** $\frac{1}{2} + \frac{1}{4} < t$ ____________

c. $y \leq 2.6 + 4.3$ ____________ **d.** $6 / g > 12$ ____________

3. Describe the solution set of each inequality.

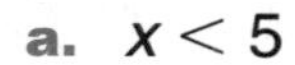

Example: $t + 5 < 8$

Solution set: All numbers less than 3

a. $8 - y > 3$ Solution set ________________________________

b. $4b \geq 8$ Solution set ________________________________

4. Graph the solution set of each inequality.

a. $x < 5$

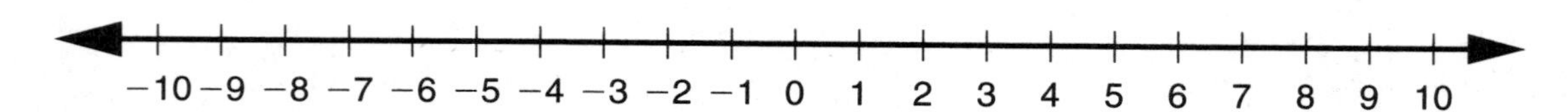

b. $6 > b$

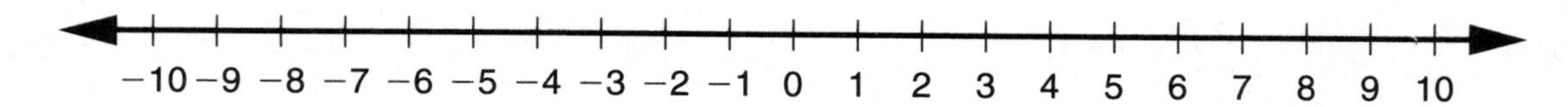

c. $1\frac{1}{2} \geq h$

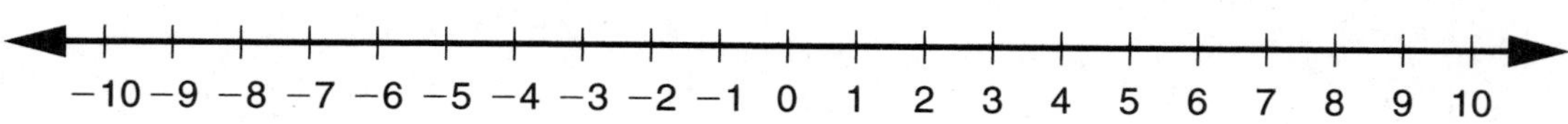

Date Time

LESSON 6•12 Math Boxes

1. Find the solution to each equation.

a. $y + 6 - 28 = 40$ $y =$ ______

b. $45 + 3n = -45$ $n =$ ______

c. $7p + 19 = -2p + 55$ $p =$ ______

d. $7 - 4w = 10w$ $w =$ ______

2. Solve the pan-balance problem.

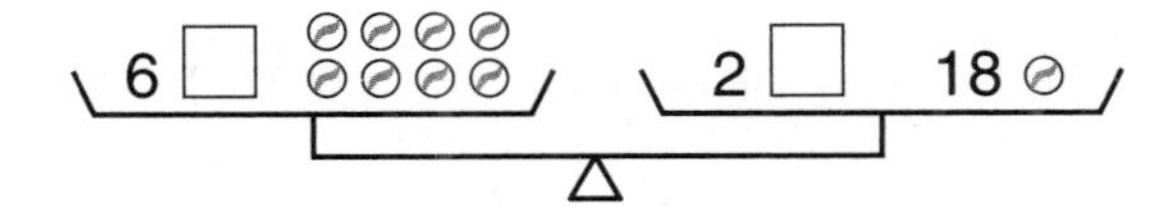

One ☐ weighs as much as ______ marbles.

3. Complete the table. Then graph the data and connect the points.

Rebecca walks at an average speed of $3\frac{1}{2}$ miles per hour.

Rule:
Distance =
$3\frac{1}{2}$ mph * number of hours

Time (hr) (*h*)	Distance (mi) ($3\frac{1}{2} * h$)
1	
2	
	$17\frac{1}{2}$
7	
	35

Rebecca's Walks

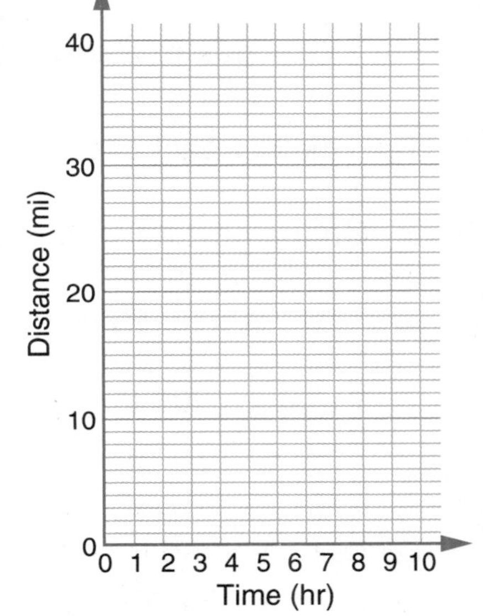

4. The spinner at the right is divided into 5 equal parts.

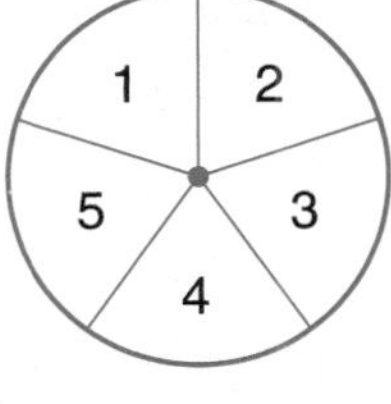

Suppose you spin this spinner 75 times. About how many times would you expect the spinner to land on a prime number?

______ times

5. Serena keeps 4 stuffed animals lined up on a shelf over her bed. How many different arrangements of the stuffed animals are possible? (*Hint:* Label the animals *A, B, C,* and *D.* Then make a list.)

______ arrangements

______

______

______

Date ______________________ Time ______________________

LESSON 6·13

Math Boxes

1. Solve. Write your answer in simplest form.

   a. $\frac{4}{52} * \frac{13}{28} =$ ______

   b. ______ $= \frac{4}{25} * \frac{75}{100}$

   c. $\frac{5}{12} + \frac{1}{4} =$ ______

   d. $\frac{3}{5} + \frac{3}{8} =$ ______

2. Rewrite each fraction as a percent.

   a. $\frac{35}{50} =$ ______

   b. $\frac{18}{24} =$ ______

   c. $\frac{7}{8} =$ ______

   d. $\frac{15}{75} =$ ______

3. The table shows the results of rolling a 6-sided die 50 times.

Number Showing	1	2	3	4	5	6
Number of Times	10	5	11	12	4	8

   Tell whether each statement below is true or false.

   a. On the next roll, a 5 is more likely to come up than a 1. ______

   b. There is a 50-50 chance of rolling a prime number. ______

   c. There is a 50-50 chance of rolling a composite number. ______

4. Cards numbered from 1 to 50 are placed in a bag. Myra picks one card without looking.

   a. What is the probability that Myra will pick a 2-digit number card?

   ______

   b. What is the probability that she will pick a number that is not a multiple of 10?

   ______

5. The manufacturer of Vita Munch cereal puts a prize in 120 boxes out of every 600.

   a. What is the probability of getting a prize if you buy a box of Vita Munch?

   ______

   b. Suppose a store has 1,800 boxes of Vita Munch in stock. About how many boxes might you expect to contain prizes?

   About ______ boxes

Date Time

Probability Concepts

Math Message

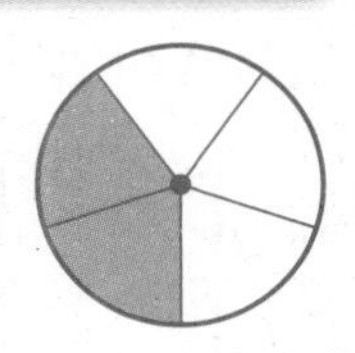

The spinner at the right has 5 equal sections. Two sections are blue. If you spin it many times, the spinner is likely to land on blue about $\frac{2}{5}$ of the time. Therefore, the probability of landing on blue is $\frac{2}{5}$, or 40%.

Using the spinners shown below, write the letter(s) of the spinner next to the statement that describes it. A spinner may be matched with more than one statement.

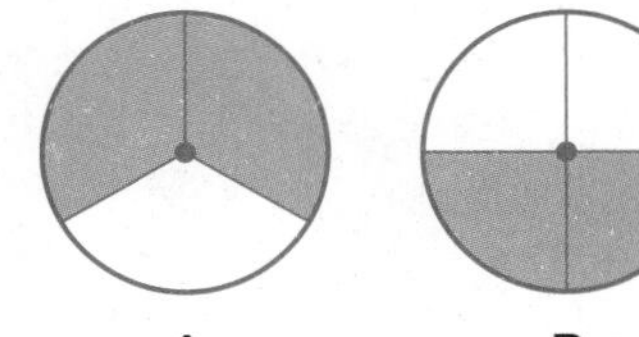
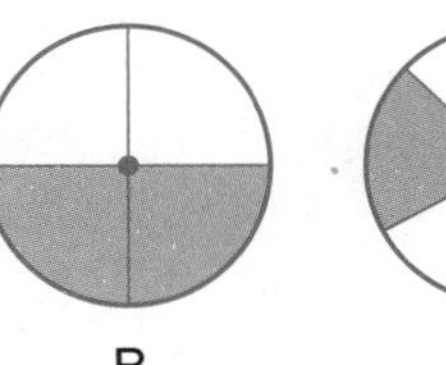
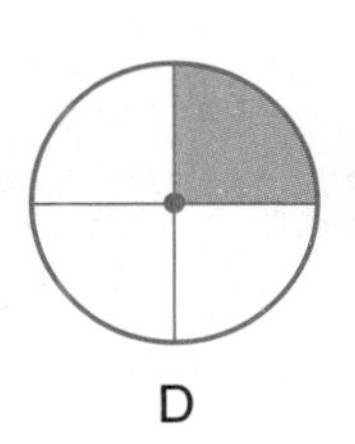
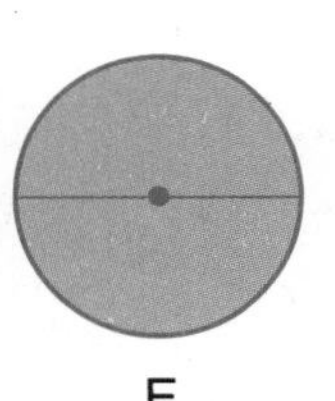
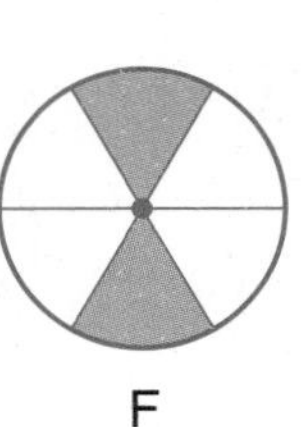

A B C D E F

Example:

This spinner will land on blue about 2 out of 3 times. A

1. There is about a $\frac{1}{4}$ chance that the spinner will land on blue. __________
2. This spinner will land on blue 100% of the time. __________
3. There is about a 50-50 chance that this spinner will land on white. __________
4. This spinner will never land on white. __________
5. The probability that this spinner will land on blue is $\frac{3}{5}$. __________
6. This spinner will land on white about twice as often as on blue. __________
7. This spinner will land on white a little less than half the time. __________
8. The probability that this spinner will land on white is 75%. __________
9. Suppose you spin Spinner A 4 times and it lands on white every time. What is the probability that the spinner will land on white on the fifth spin? __________
10. If you spin Spinner A 90 times, how many times would you expect the spinner to land on blue? __________

Date Time

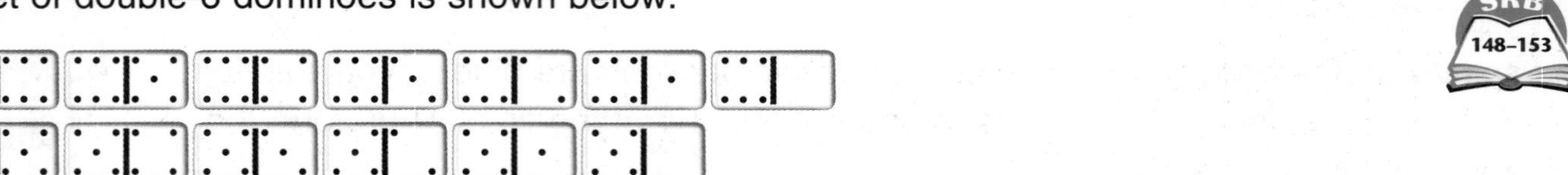

LESSON 7•1 Domino Probabilities

A set of double-6 dominoes is shown below.

SRB 148–153

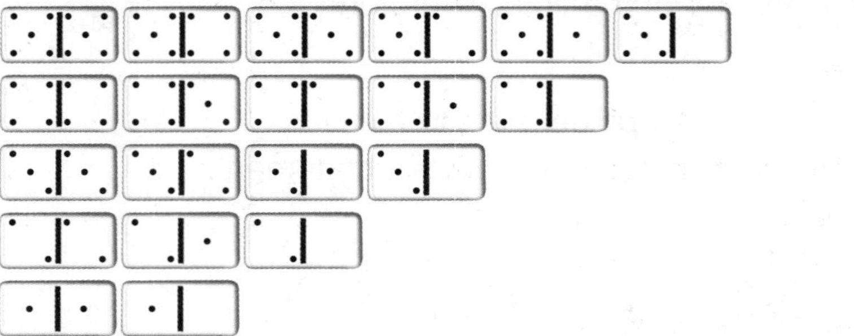

Suppose all the dominoes are turned facedown and mixed thoroughly.
You select one domino and turn it faceup.

1. How many possible outcomes are there? ________ possible outcomes

2. Are the outcomes equally likely? ________

When the possible outcomes are equally likely, the following formula is used to find the probabilities:

A favorable outcome is the outcome that makes an event happen.

$$\text{Probability of an event} = \frac{\text{number of favorable outcomes}}{\text{number of possible outcomes}}$$

3. What is the probability of selecting each domino? ________

Find the probability of selecting each domino described below.

4. A double ________

5. Exactly one blank side ________

6. No blank sides ________

7. The sum of the dots is 7. ________

8. The sum of the dots is greater than 7. ________

9. Exactly one side is a 3. ________

10. Both sides are odd numbers. ________

Date Time

LESSON 7•1

Math Boxes

1. Fill in the missing equivalents. Write fractions in simplest form.

Fraction	Decimal	Percent
$\frac{9}{10}$		
	0.98	
		60%
$\frac{7}{25}$		
		12.5%

2. Solve the following equations. Check each solution by substituting it for the variable in the original equation.

a. $8x - 1 = 11$

Solution: $x =$ ________

b. $\frac{2}{5}y + 6 = 10$

Solution: $y =$ ________

c. $-20m + 20 = -20$

Solution: $m =$ ________

3. Indicate whether each inequality is true or false.

a. $\frac{2}{3} * 9 > 8$ ________

b. $-4 \leq -3 - 1$ ________

c. $48 - (6 * 4) > 20$ ________

d. $8 - 10 \neq 13 - 15$ ________

4. Graph the solution set for $k > -2$ on the number line below.

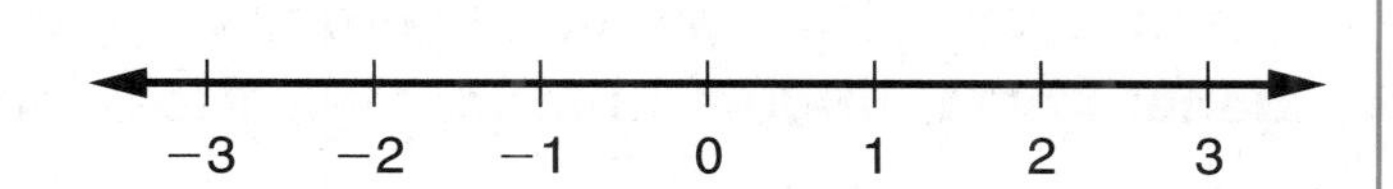

5. Complete.

a. ________ m = 368 mm

b. ________ cm = 0.245 m

c. 32 mm = ________ m

d. 45.2 cm = ________ mm

e. 0.25 mm = ________ cm

6. Solve.

Solution

a. $\frac{w}{8} = 16$ ________

b. $\frac{60}{p} = 5$ ________

c. $\frac{3}{7} = \frac{t}{28}$ ________

d. $\frac{d}{18} = \frac{4}{6}$ ________

Date Time

Generating Random Numbers

Math Message

Suppose you have a deck of number cards, one card for each of the numbers 1 through 5. When you shuffle the cards and pick a card without looking, the possible outcomes are 1, 2, 3, 4, and 5. These outcomes are equally likely. Numbers found in this way are **random numbers.**

Suppose you continue finding random numbers using these steps.

- Shuffle
- Pick a card
- Replace the card
- Repeat

1. If you did this many times, about what percent of the time would you expect to pick the number 5? About ________ percent

An Experiment

2. Work with a partner in a group of 4 students. Use a deck of 5 number cards, one card for each of the numbers 1 through 5.

3. One of you shuffles the deck of 5 number cards and fans them out facedown. Your partner then picks one without looking. The pick **generates a random number** from 1 to 5. The number is an **outcome.**

4. The person picking the card tallies the outcome in the table below while the person with the deck replaces the card and shuffles the deck. Generate exactly 25 random numbers.

Outcome	Tally	Number of Times Picked
1		
2		
3		
4		
5		
	Total Random Numbers	**25**

LESSON 7•2

Generating Random Numbers *continued*

5. Record the results in the table below. In the My Partnership column, write the number of times each of the numbers 1 through 5 appeared.

6. In the Other Partnership column, record the results of the other partnership in your group.

7. For each outcome, add the two results and write the sum in the Both Partnerships column.

8. Convert each result in the Both Partnerships column to a percent and record it in the % of Total column. For example, 10 out of 50 would be 20%.

Outcome	My Partnership	Other Partnership	Both Partnerships	% of Total
1			_____ out of 50	
2			_____ out of 50	
3			_____ out of 50	
4			_____ out of 50	
5			_____ out of 50	
Total	**25**	**25**	**50 out of 50**	**100%**

Date Time

LESSON 7•2

More Probability

1. The table below shows the results of a survey of 200 health club members. The members were asked to name the area of the club in which they did the most exercise.

Exercise Area	Number of People
Swimming pool	80
Aerobics studio	28
Racquetball court	70
Weight room	22

What is the probability, written as a percent, that a member spends most of his or her exercise time in

a. the aerobics studio? ______________

b. the swimming pool? ______________

c. an area that is *not* the racquetball court? ______________

d. the weight room or the racquetball court? ______________

2. Suppose the value of x is chosen at random from the following set of numbers: {1, 2, 3, 4, 5}.

a. What is the probability that the area of the rectangle at the right is greater than 65 in.^{2}? Express the probability as a percent.

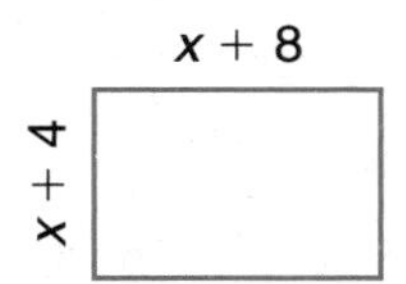

____

b. What is the probability that the perimeter of the rectangle is less than 25 in.? Express the probability as a percent.

____

Date Time

LESSON 7•2

Math Boxes

1. Suppose one of the squares is chosen at random. What is the probability that the square is

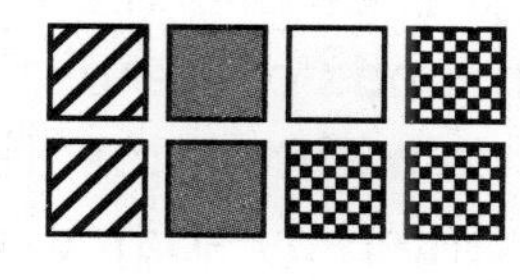

a. white? ________

b. checkered? ________

c. checkered or white? ________

2. Complete.

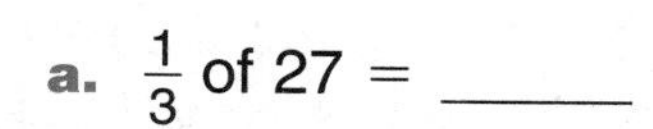

a. $\frac{1}{3}$ of 27 = ______

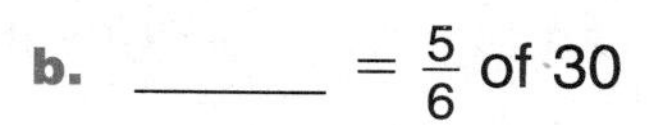

b. ______ = $\frac{5}{6}$ of 30

c. ______ = $\frac{4}{7}$ of 42

d. $\frac{3}{8}$ of 56 = ______

3. Complete the table for the formula below. Then plot the points to make a graph.

Formula: $2s - 5 = t$

s	*t*
1	
2	
5	
	11
	15

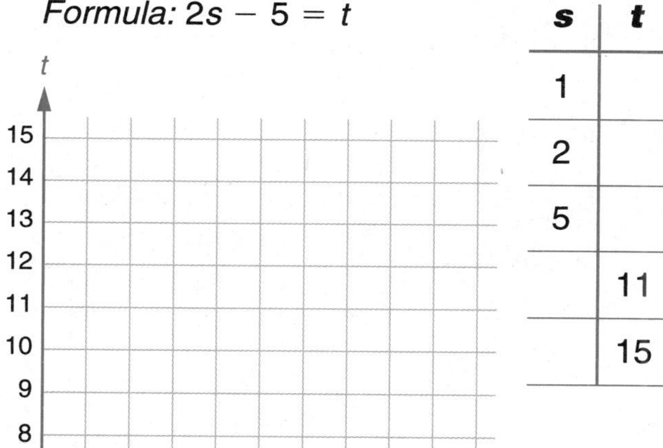

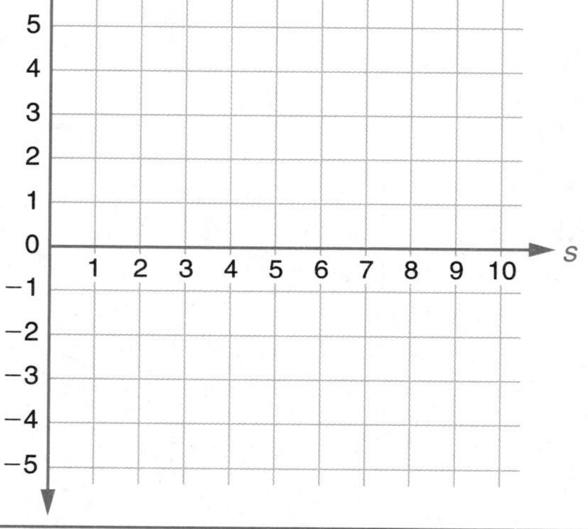

4. Name two solutions for each inequality.

Solutions

a. $1.1 - 0.38 < w$ ________________

b. $10\frac{1}{3} - 6\frac{2}{3} \geq g$ ________________

5. Write each ratio in simplest form.

a. 8 to 10 ________

b. 35 out of 100 ________

c. 60 wins to 90 losses ________

d. \$144 in 12 hours ________

SRB 117–119

Date Time

LESSON 7•3

Using Random Numbers

Suppose two evenly matched teams play a fair game that cannot end in a tie—one team must win. Because the teams have an equal chance of winning, you could get about the same results by tossing a coin. If the coin lands on HEADS, Team 1 wins; if it lands on TAILS, Team 2 wins. In this way, tossing a coin **simulates** the outcome of the game. In a **simulation,** an object or event is represented by something else.

Suppose Team 1 and Team 2 play a best-of-5 tournament. The first team to win 3 games wins the tournament. Use a coin to simulate the tournament as follows:

Games 1–3 If the coin lands on HEADS, Team 1 wins. If the coin lands on TAILS, Team 2 wins.

Games 4 and 5 Play only if necessary. Repeat the instructions for Game 1.

Sample results:

If the coin tosses are HEADS, HEADS, HEADS, Team 1 wins the tournament.

If the coin tosses are HEADS, HEADS, TAILS, HEADS, Team 1 wins.

If the coin tosses are TAILS, HEADS, HEADS, TAILS, TAILS, Team 2 wins.

1. Fill in the table as described on the next page.

Number of Games Needed to Win the Tournament	Winner	Tally of Tournaments Won	Total Tournaments Won
3	Team 1		
	Team 2		
4	Team 1		
	Team 2		
5	Team 1		
	Team 2		
		Total	**25**

LESSON 7•3

Using Random Numbers *continued*

2. Use coin tosses to play a best-of-5 tournament. Make a tally mark in the Tally of Tournaments Won column of the table on page 254. The tally mark shows which team won the tournament and in how many games.

3. Play exactly 24 more tournaments. Make a tally mark to record the result for each tournament. Then convert the tally marks into numbers in the Total Tournaments Won column.

4. Use the table on page 254 to estimate the chance that a tournament takes

   a. exactly 3 games. ________%

   b. exactly 4 games. ________%

   c. exactly 5 games. ________%

   d. fewer than 5 games. ________%

Discuss the following situations with a partner. Record your ideas.

5. Suppose there is a list of jobs that need to be done for your class (such as distributing supplies, collecting books, and taking messages to the office). How might you use random numbers to assign the jobs, without using any pattern or showing favoritism?

____

____

____

____

6. You want to play a game. The directions are: *Roll 2 dice and add the numbers. Move your marker ahead that many spaces.* You do not have any dice. How can you use number cards to play the game?

____

____

____

____

LESSON 7·3

Math Boxes

1. Fill in the missing equivalents. Write fractions in simplest form.

Fraction	Decimal	Percent
$\frac{7}{8}$		
	$0.\overline{6}$	
$\frac{7}{10}$		
		62.5%
	0.005	

2. Solve the following equations. Check each solution by substituting it for the variable in the original equation.

a. $k = 8k + 28$

Solution: $k =$ ________

b. $20n - 28 = 10n + 2$

Solution: $n =$ ________

c. $-p - 1 = p - 21$

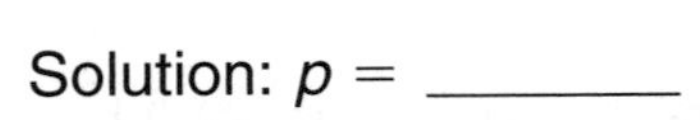
Solution: $p =$ ________

3. Which inequality is false when $a = 3$, $b = -2$, and $c = 6$?
Fill in the circle next to the best answer.

Ⓐ $b^2 - 4ac < 0$

Ⓑ $0 < \frac{c}{a} + b$

Ⓒ $a(b + c) \geq ab + ac$

Ⓓ 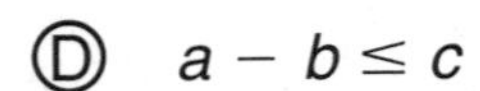 $a - b \leq c$

4. Graph the solution set for $b \neq 0$ on the number line below.

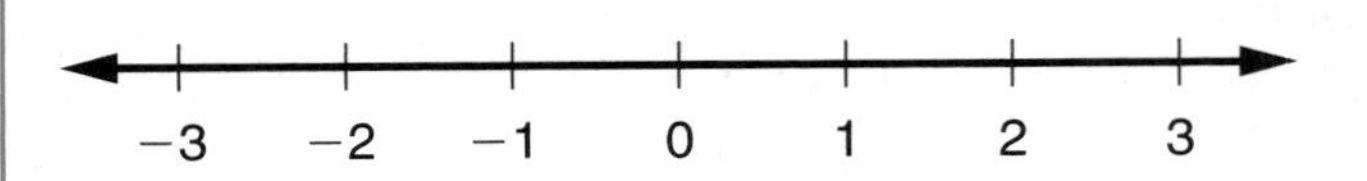

5. Complete.

a. 15 m = ____________ cm

b. 25 cm = ____________ m

c. 143 mm = ____________ cm

d. 2.06 cm = ____________ mm

e. ____________ mm = 1.43 cm

6. Solve.

Solution

a. $\frac{n}{6} = 4$ ________________

b. $\frac{42}{b} = 6$ ________________

c. $\frac{5}{8} = \frac{g}{32}$ ________________

d. $\frac{k}{21} = \frac{2}{14}$ ________________

Date Time

Mazes and Tree Diagrams

Math Message

The diagram at the right shows a maze. A person walking through the maze does not know in advance how many paths there are or how they divide.

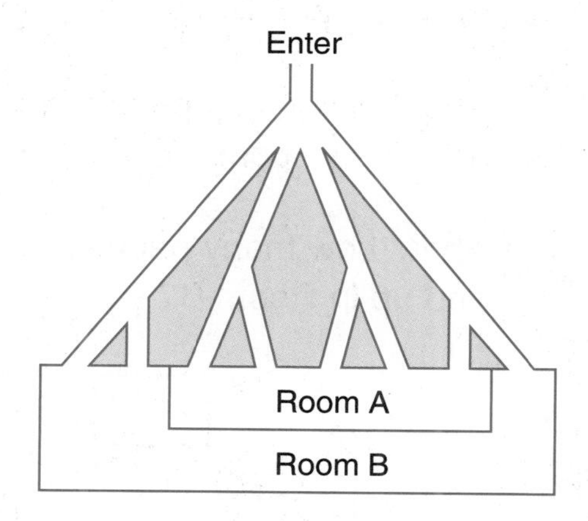

Pretend that you are walking through the maze. Each time the path divides, you select your next path at random. Each path at an intersection has the same chance of being selected. You may not retrace your steps.

Depending on which paths you follow, you will end up in Room A or Room B.

1. In which room are you more likely to end up—Room A or Room B? ____________

2. Suppose 80 people took turns walking through the maze.

   a. About how many people would you expect to end up in Room A? ____________

   b. About how many people would you expect to end up in Room B? ____________

Your teacher will show you how to complete the following tree diagram. You can read about tree diagrams on pages 154 and 155 in the *Student Reference Book.*

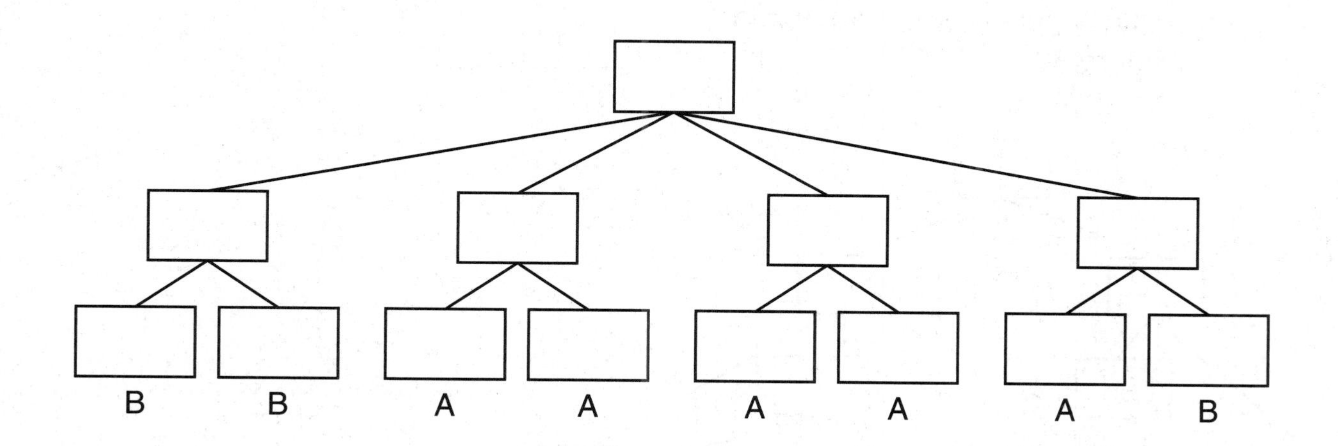

Date Time

LESSON 7·4

Maze Problems

1. Use the tree diagram below to help you solve the following problem. Suppose 60 people walk through the maze below.

   a. About how many people would you expect to end up in Room A? ______________

   b. About how many people would you expect to end up in Room B?

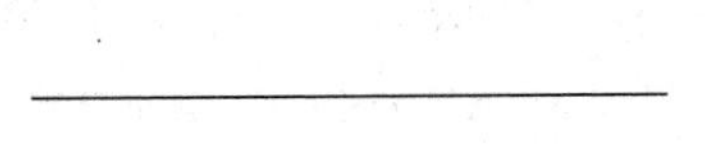

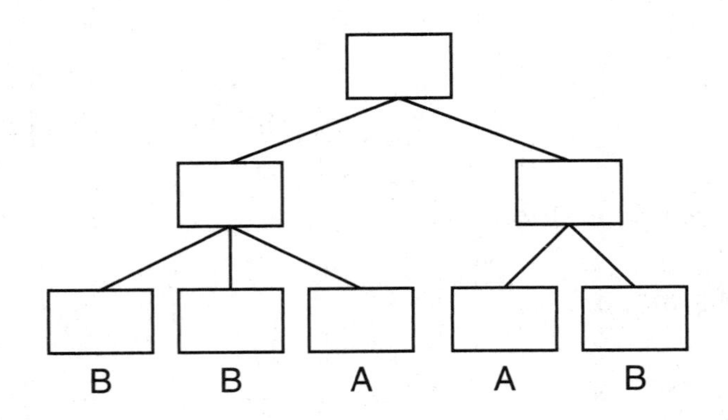

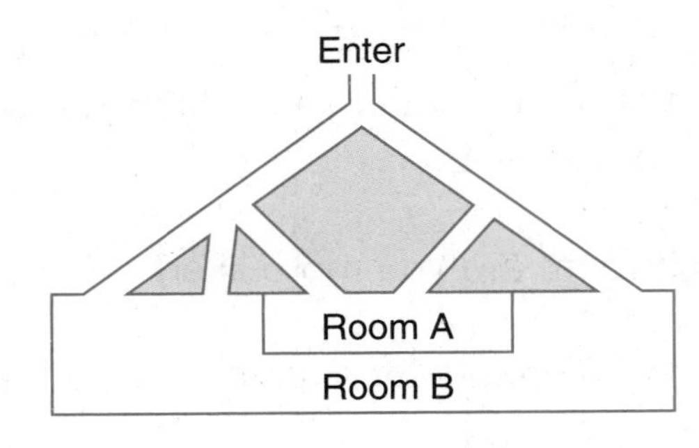

2. Make a tree diagram to help you solve the following problem. Suppose 120 people walk through the maze below.

   a. About how many people would you expect to end up in Room A?

   b. About how many people would you expect to end up in Room B?

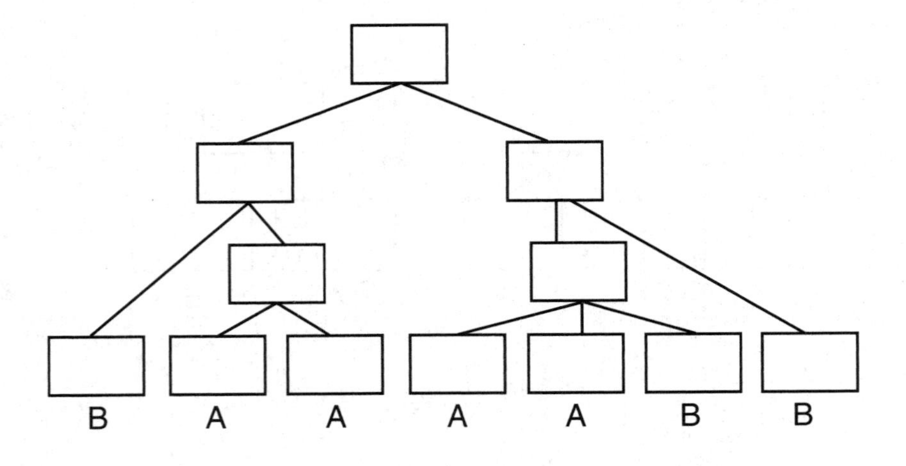

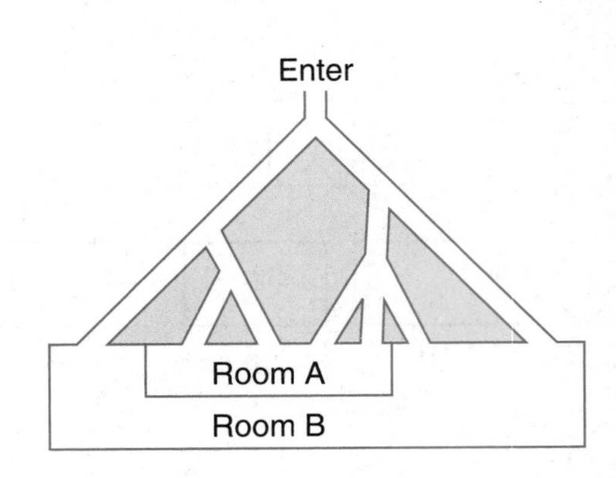

Date Time

LESSON 7•4

Math Boxes

1. Suppose one of the numbered marbles is chosen at random.

What is the probability of choosing

a. a number ≤ 5? ______ fraction ______ percent

b. a 3 or an 8? ______ fraction ______ percent

c. a prime or a composite number? ______ fraction ______ percent

2. Complete.

a. $\frac{3}{4}$ of $80 =$ ______

b. ______ $= \frac{2}{9}$ of 27

c. ______ $= \frac{4}{5}$ of 60

d. $\frac{7}{11}$ of $88 =$ ______

3. Complete the table for the formula below. Then plot the points to make a graph.

Formula: $4h = g$

h	*g*
1	
2	
3	
	20
	26

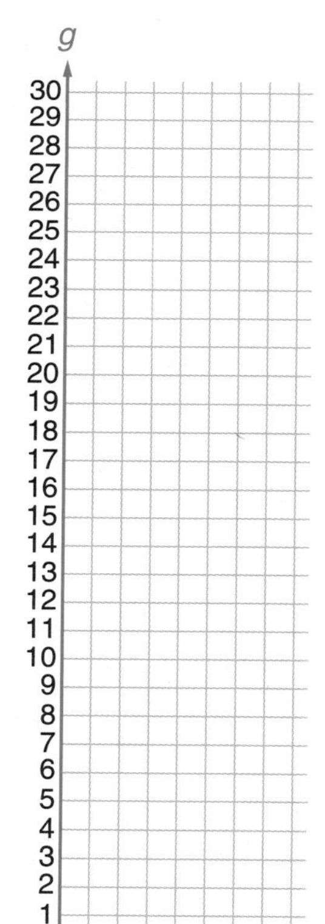

4. Name two solutions for each inequality.

Solutions

a. $n \geq 3\frac{11}{16} + 4\frac{1}{2}$ ______

b. $k < 6\frac{2}{5} * 15$ ______

5. Write each ratio in simplest form.

a. 20 to 12 ______

b. 64 out of 78 ______

c. 14 boys to 16 girls ______

d. 440 miles in 8 hours ______

SRB
117–119

Date Time

LESSON 7•5

Probability Tree Diagrams

Complete the tree diagram for each maze.

Write a fraction next to each branch to show the probability of selecting that branch. Then calculate the probability of reaching each endpoint. Record your answers in the blank spaces beneath the endpoints.

1. What is the probability of entering Room A? ______

What is the probability of entering Room B? ______

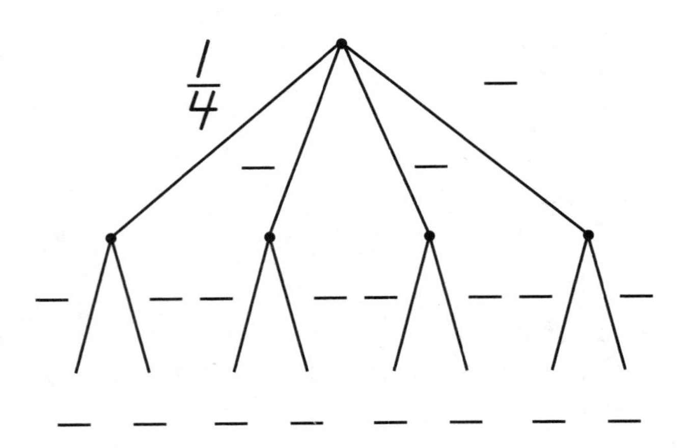

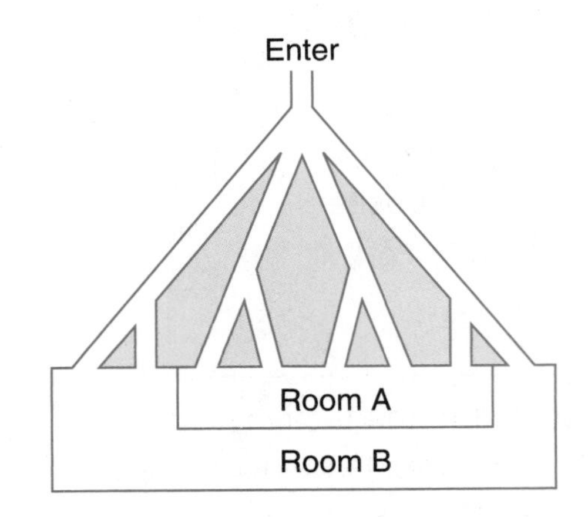

2. What is the probability of entering Room A? ______

What is the probability of entering Room B? ______

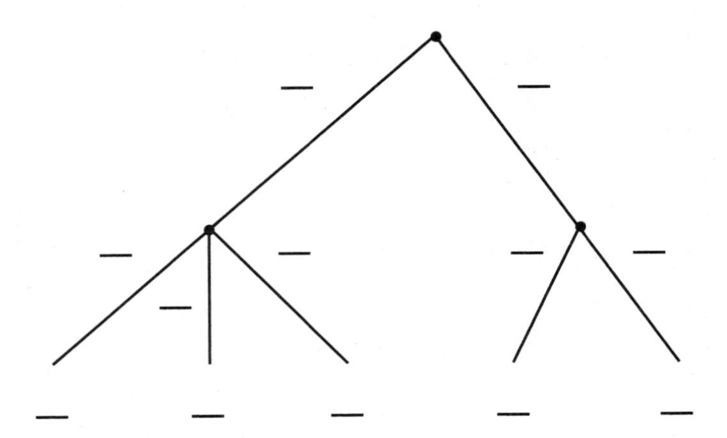

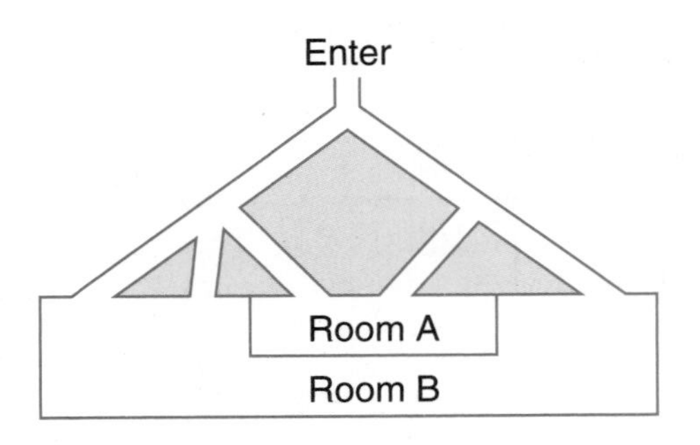

Date Time

LESSON 7•5 Probability Tree Diagrams *continued*

3. Josh has 3 clean shirts (red, blue, and green) and 2 clean pairs of pants (tan and black). He randomly selects one shirt. Then he randomly selects a pair of pants.

a. Complete the tree diagram by writing a fraction next to each branch to show the probability of selecting that branch. Then calculate the probability of selecting each combination.

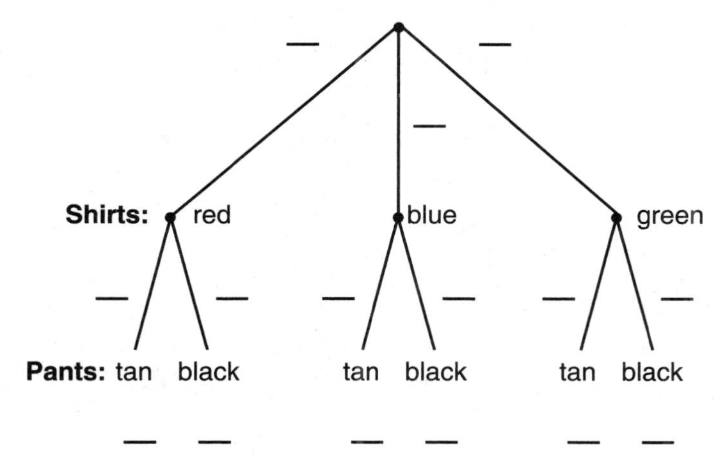

b. List all possible shirt-pants combinations. One has been done for you.

red–tan ______________________________

c. How many different shirt-pants combinations are there? ____________

d. Do all the shirt-pants combinations have the same chance of being selected? ________

e. What is the probability that Josh will select

the blue shirt? ________ the blue shirt and the tan pants? ________

the tan pants? ________ a shirt that is not red? ________

the black pants and a shirt that is not red? ________

Try This

Suppose Josh has 4 clean shirts and 3 clean pairs of pants. Explain how to calculate the number of different shirt-pants combinations without drawing a tree diagram.

____

____

Date Time

LESSON 7•5 Math Boxes

1. Darnell has 3 jackets and 4 baseball hats. Complete the tree diagram.

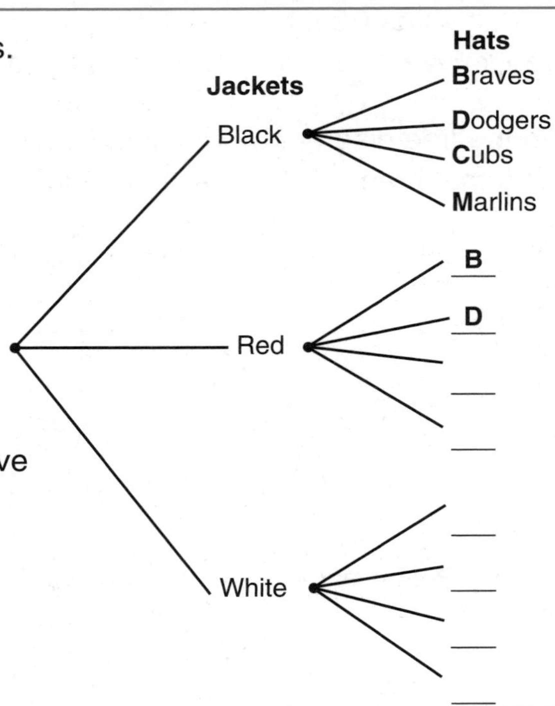

a. How many jacket-hat combinations are possible?

____

b. Do all the jacket-hat combinations have the same chance of being selected?

____

SRB 155 156

2. Which set of numbers is represented by the graph below? Choose the best answer.

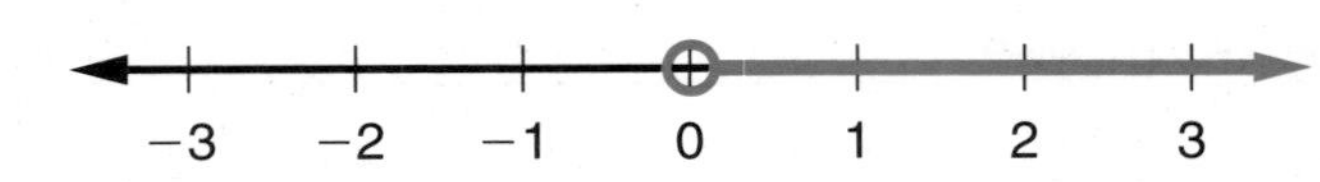

- ◯ positive real numbers
- ◯ positive integers
- ◯ whole numbers
- ◯ counting numbers

3. Write each number in standard notation.

a. 72 billion _______________

b. 42.78 million _______________

c. 89.6 billion _______________

d. 0.5 million _______________

4. Janella walks at a speed of 6.9 kilometers per hour. At this rate, how far can she walk

a. in 2 hours? _______ kilometers

b. in 20 minutes? _______ kilometers

c. in 1 hour 40 minutes?

_______ kilometers

5. Express the length of $\overline{JK}$ to the length of $\overline{JL}$ as a simplified fraction.

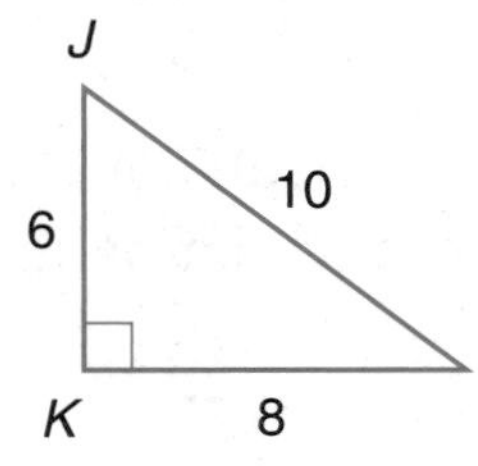

$\frac{JK}{JL}$ = _______

Date Time

LESSON 7·6

Math Boxes

1. The Venn diagram shows the relationships between sets of numbers within the set of real numbers. Use the diagram to answer the following questions.

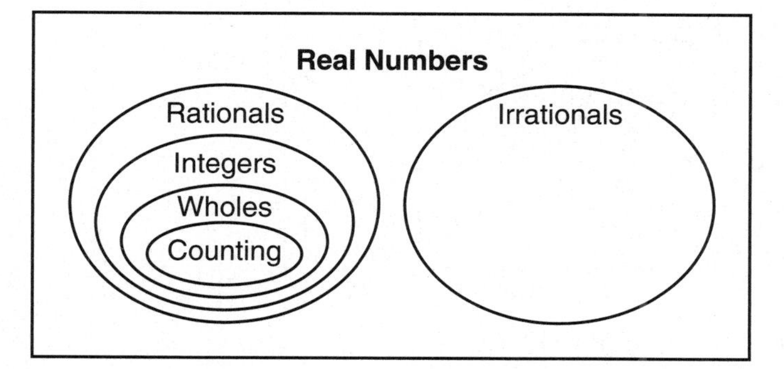

a. Can a number be rational and irrational?

____

b. Is a whole number a rational number?

____

2. Complete.

a. $\frac{1}{10}$ of 268 = __________

b. $\frac{1}{100}$ of 21,509 = __________

c. $\frac{1}{1,000}$ of 7,834 = __________

d. $\frac{1}{100}$ of 72 = __________

3. Fill in the missing numbers.

a. 947 * 23 * 16 = 16 * 23 * __________

b. 18 * 7 * 3 = 21 * __________

c. __________ * 51 * 97 = 51 * 97 * 82

d. __________ * 14 * 182 = 28 * 182

e. __________ * 29 * 30 = 150 * 29

4. Use division to rename the fraction as a decimal rounded to the nearest hundredth. Show your work.

$\frac{14}{15}$

$\frac{14}{15}$ = __________

5. Estimate the percent equivalents for the following fractions.

a. $\frac{3}{26}$ Estimate __________

b. $\frac{7}{9}$ Estimate __________

c. $1\frac{3}{7}$ Estimate __________

Date Time

Venn Diagrams

Math Message

The Venn diagram below shows the factors of 20 and 30.

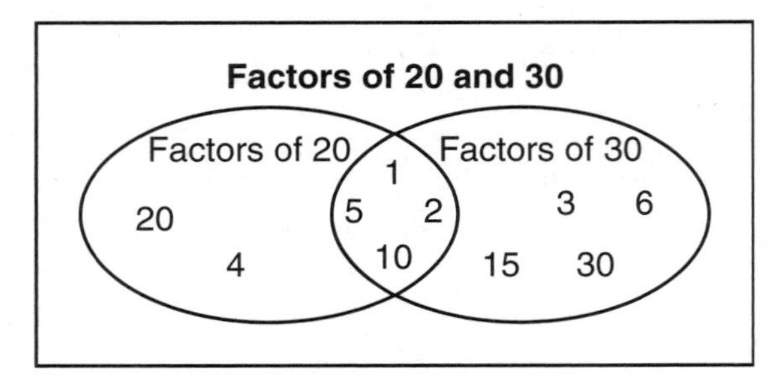

1. Which numbers are factors of 20 but are *not* factors of 30? ____________

2. Which numbers are factors of both 20 and 30? ____________

3. List the factors of 30. ____________

4. What is the greatest common factor of 20 and 30? ____________

---

5. Ms. Barrie teaches math and science. The Venn diagram below shows the number of students in her classes.

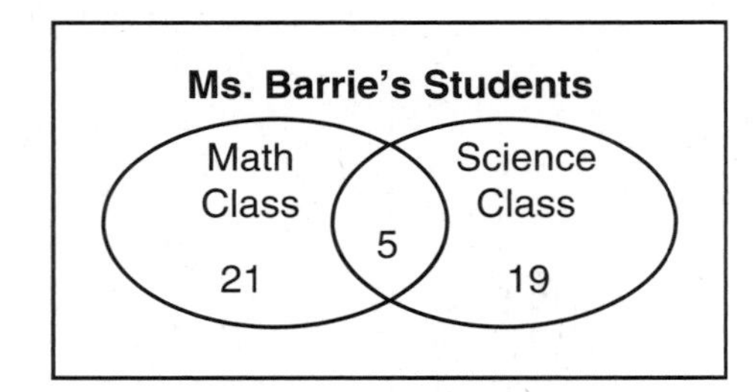

   a. How many students are in Ms. Barrie's math class? ____________ students

   b. How many students are in her science class? ____________ students

   c. How many students have Ms. Barrie for math and science? ____________ students

   d. How many students have Ms. Barrie as a teacher in at least one class? ____________ students

Date Time

LESSON 7·6 Venn Diagrams *continued*

6. The sixth graders at Lincoln Middle School were asked whether they write with their left or right hand. A small number of students reported that they were *ambidextrous,* which means they write equally well with either hand.

The survey results are shown in the Venn diagram at the right.

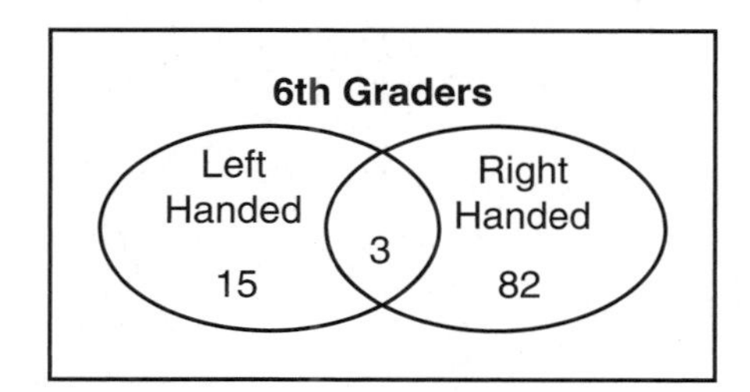

a. How many students were surveyed? _______ students

b. How many students are ambidextrous (can write with either hand)? _______ students

c. How many students always write with their left hand? _______ students

d. How many students never write with their left hand? _______ students

7. Mr. Wu has 32 students in his sixth-grade homeroom. He identified all students who scored 90% or above on each test. Mr. Wu then drew the Venn diagram at the right.

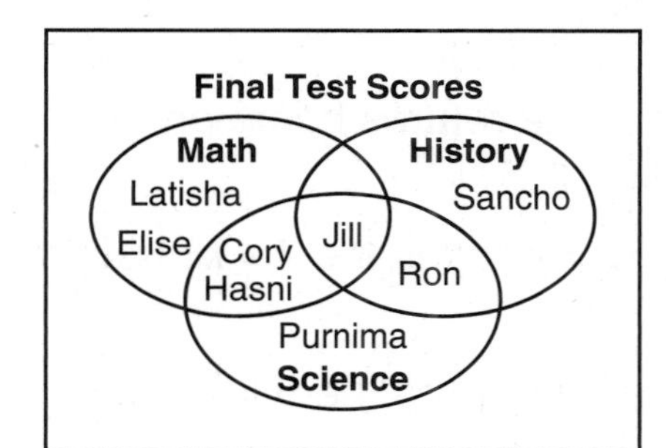

a. Which student(s) scored 90% or above on *all* 3 tests?

____

b. Which student(s) scored 90% or above on *exactly* 1 test?

____

c. Which student(s) scored 90% or above in *both* math and science?

____

Try This

d. What percent of the students in Mr. Wu's homeroom had a score of 90% or above on *at least* 2 tests? _______

Date Time

LESSON 7·6 Probability Tree Diagrams

Mr. Gulliver travels to and from work by train. Trains to work leave at 6, 7, 8, 9, and 10 A.M. Trains from work leave at 3, 4, and 5 P.M. Suppose Mr. Gulliver selects a morning train at random and then selects an afternoon train at random.

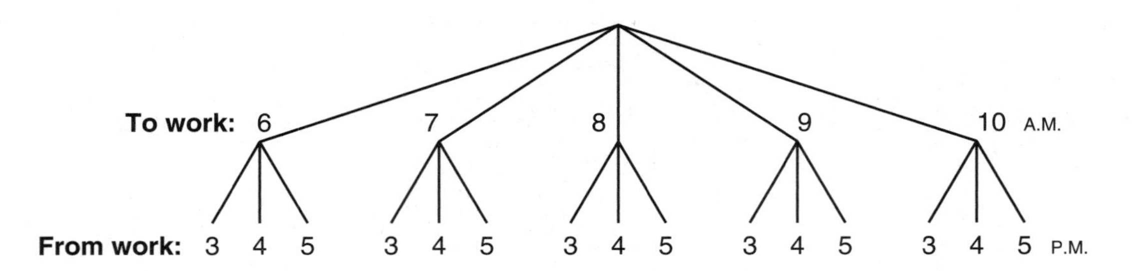

1. How many different combinations of trains to and from work can Mr. Gulliver take?

   ________ combinations

Calculate the probability of each of the following.

2. Mr. Gulliver takes the 7 A.M. train to work. ____________
3. He returns home on the 4 P.M. train. ____________
4. He takes the 7 A.M. train to work and returns home on the 4 P.M. train. ____________
5. He takes the 9 A.M. train to work and returns home on the 5 P.M. train. ____________
6. He leaves for work *before* 9 A.M. ____________
7. He leaves for work at 6 A.M. or 7 A.M. and returns home at 3 P.M. ____________
8. He returns home, but not on the 5 P.M. train. ____________
9. He boards the return train 9 hours after leaving for work. ____________

Date Time

LESSON 7•7

Math Boxes

1. The coin is flipped, and the spinner is spun. Complete the tree diagram. Write a fraction next to each branch to show the probability of selecting that branch.

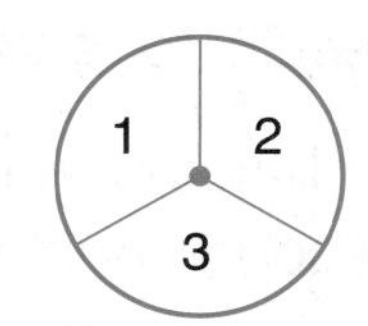

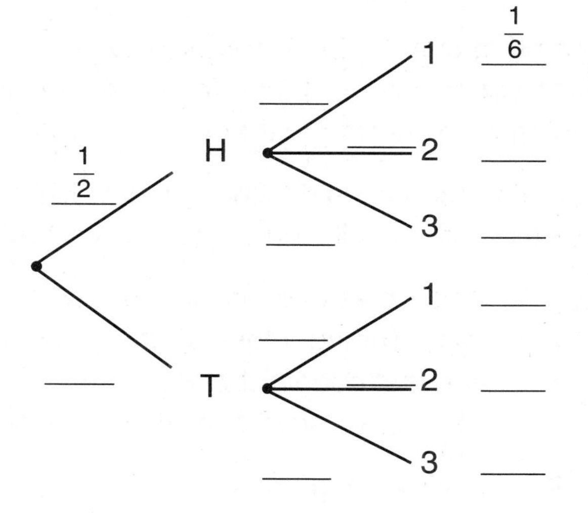

How many different outcomes are possible? ________

2. Describe the solution set for the graph below.

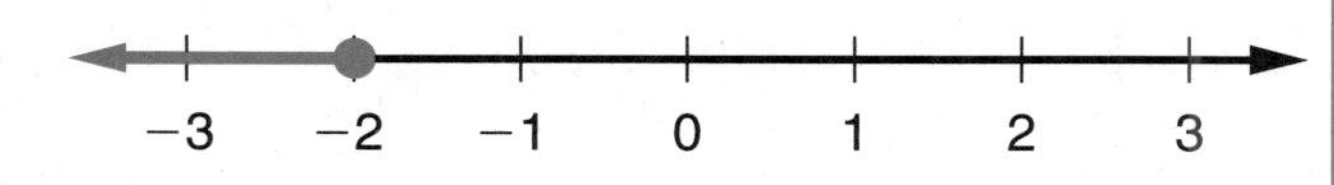

Solution set:

____

3. Write each number in standard notation.

a. 14.05 billion ________________

b. 2.01 trillion ________________

c. 0.25 million ________________

d. 0.75 thousand ________________

4. Suppose the average dalmation dog runs 1,668 feet in 1 minute. At this rate, how far can this dog run

a. in 3 minutes? ________ feet

b. in 1 minute 30 seconds?

________ feet

c. in 45 seconds? ________ feet

5. Express the ratio of the perimeter of square *ABCD* to the perimeter of square *EFGH* as a fraction in simplest form.

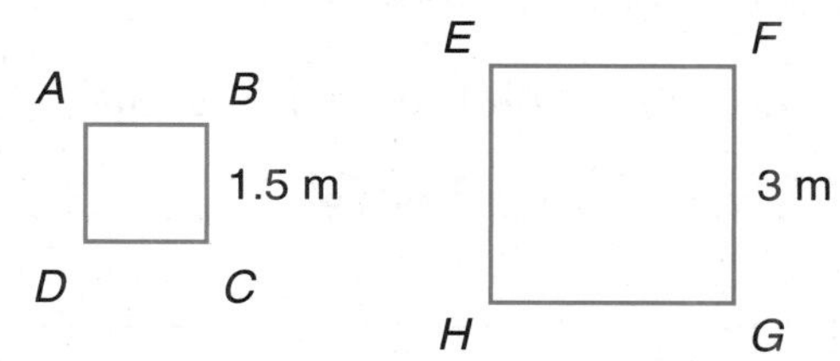

$\frac{\text{Perimeter } ABCD}{\text{Perimeter } EFGH} =$ ________

Date _______________ Time _______________

LESSON 7·7

Fair and Unfair Games

Math Message

A game of chance for 2 or more players is a **fair game** if each player has the same chance of winning. A game for 1 player is fair if the player has an equal chance of winning or losing. Any other game is an **unfair game.**

Each of the 4 games described below is for 1 player. Play Games 1, 2, and 3 a total of 6 times each. Tally the results. Later, the class will combine results for each game.

Game 1 Put 2 black counters and 1 white counter into a paper bag and shake the bag. Without looking, draw 1 counter. Then draw a second counter without putting the first counter back into the bag. If the 2 counters are the same color, you win. Otherwise, you lose. Play 6 games and tally your results.

● ●
○

Tally for 6 games: Win _______________ Lose _______________

Do you think the game is fair? _______________

Combined class data: Win _______________ Lose _______________

Game 2 Use 2 black counters and 2 white counters. The rules are the same.

● ●
○ ○

Tally for 6 games: Win _______________ Lose _______________

Do you think the game is fair? _______________

Combined class data: Win _______________ Lose _______________

Game 3 Use 3 black counters and 1 white counter. The rules are the same.

● ●
● ○

Tally for 6 games: Win _______________ Lose _______________

Do you think the game is fair? _______________

Combined class data: Win _______________ Lose _______________

Game 4 Suppose you use 4 black counters. The rules are the same.

● ●
● ●

Do you think the game is fair? _______________

Explain your answer. ____

____

____

Date Time

Fair Games and Probability

You can use a tree diagram to decide whether a game is fair or unfair. This tree diagram represents Game 1 on page 268.

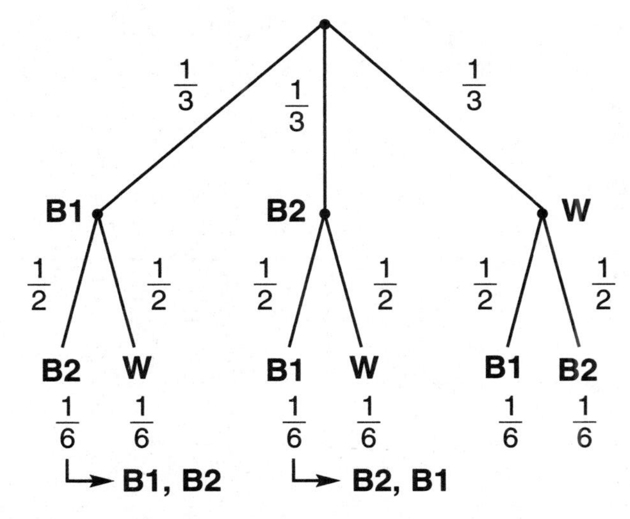

Before you draw the first counter, there are 3 counters in the bag. Although the 2 black counters look alike, they are not the same. To tell them apart, they are labeled B1 and B2. The probability of drawing either B1, B2, or W is $\frac{1}{3}$.

After the first draw, there are 2 counters left in the bag. The probability of drawing either of the 2 remaining counters is $\frac{1}{2}$.

There are 6 possible ways to draw the 2 remaining counters. The probability of each outcome is $\frac{1}{3} * \frac{1}{2}$, or $\frac{1}{6}$. There are 2 ways to draw the same color counters:

Draw B1 on the first draw and B2 on the second draw.
Draw B2 on the first draw and B1 on the second draw.

The chance of drawing 2 black counters is $\frac{1}{6} + \frac{1}{6}$, which is $\frac{2}{6}$, or $\frac{1}{3}$. Therefore, Game 1 is not a fair game.

Make tree diagrams to help you answer these questions.

1. What is the probability of winning Game 2? ____________

2. Is Game 2 a fair game? ____________

3. What is the probability of winning Game 3? ____________

4. Is Game 3 a fair game? ____________

Date Time

LESSON 7•8

Probabilities and Outcomes

Math Message

1. A game is played using the spinner at the right. The spinner will land on white $\frac{4}{5}$ of the time. Each time the spinner lands on white, Player A gets 1 point. Each time the spinner lands on blue, Player B gets 4 points. The winner is the player with more points after 20 spins.

   Is this a fair game? Explain. ______________________________

   ____
   ____
   ____
   ____
   ____
   ____
   ____

2. Suppose you are taking a multiple-choice test. Four possible answers are given for each question. There are 20 questions for which you don't know the correct answer. You decide to guess the answer for each of these questions.

   a. What is the probability of answering a question correctly? ______

   b. How many of the 20 questions would you expect to answer correctly by guessing? ______ questions

   Explain. ______________________________

3. In scoring the test, each correct answer is worth 1 point. To discourage guessing, there is a penalty of $\frac{1}{3}$ point for each incorrect answer. Do you think this is a fair penalty?

   Explain. ______________________________

   ____
   ____
   ____
   ____

Date Time

LESSON 7·8 Math Boxes

1. The Venn diagram shows the relationships between the family of quadrangles that includes squares, rhombuses, rectangles, trapezoids, and parallelograms. Use the diagram to indicate whether each statement is true or false.

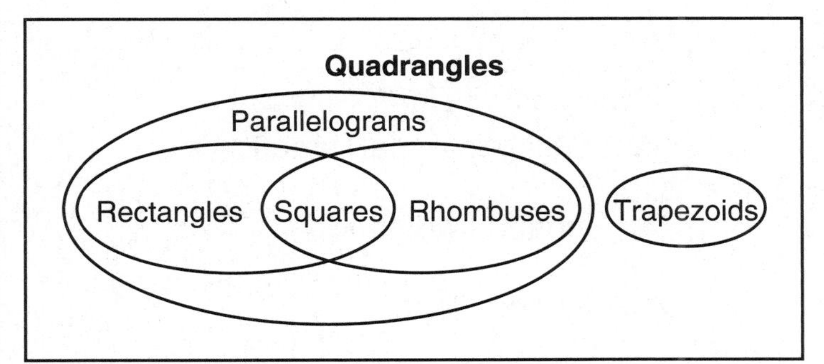

a. A square is a rectangle. __________

b. A rhombus is always a square. __________

c. A trapezoid is a quadrangle. __________

2. Write each fraction as a decimal.

a. $\frac{709}{10}$ = __________

b. $\frac{83,261}{1,000}$ = __________

c. $\frac{352}{10,000}$ = __________

d. $\frac{247.8}{100}$ = __________

3. Fill in the missing numbers.

a. 58 * 91 * 27 = 27 * 58 * __________

b. 24 * 16 * 10 = 240 * __________

c. __________ * 500 = 25 * 20 * 153

d. __________ * 426 * 81 = 81 * 945 * 426

e. __________ * 35 * 94 = 35 * 94 * 87

4. Use division to rename the fraction as a decimal rounded to the nearest hundredth. Show your work.

$\frac{31}{36}$

$\frac{31}{36}$ = __________

5. Which is the best estimate for the percent equivalent of $\frac{3}{40}$?

Circle the best answer.

A. 3%

B. 8%

C. 10%

D. 13%

Date Time

LESSON 7·8

Guessing on Multiple-Choice Tests

1. For each question on the following test, first draw a line through each answer that you know is not correct. Then circle one answer for each question. If you do not know the correct answer, guess.

Number correct ______________

1. A nautical mile is equal to

   A. 1 foot.

   B. 1 yard.

   C. 1,832 meters.

   D. 1,852 meters.

2. In 2000, the population of Nevada was

   A. 1 billion.

   B. 2,810,828.

   C. 2,018,828.

   D. 14,628.

3. Which region receives the greatest average annual rainfall?

   A. Atlanta, Georgia

   B. New Orleans, Louisiana

   C. Mojave Desert, California

   D. Sahara Desert, Africa

4. The addition sign (+) was introduced into mathematics by

   A. Johann Widman.

   B. Johann Rahn.

   C. Abraham Lincoln.

   D. Martin Luther King, Jr.

5. How many diagonals does a 13-sided polygon (13-gon) have?

   A. 1

   B. 54

   C. 65

   D. 13,000

6. The leading cause of death in the United States is

   A. bungee jumping.

   B. cancer.

   C. drowning.

   D. heart disease.

Date Time

LESSON 7·8

Guessing on Multiple-Choice Tests *continued*

2. When you can narrow the choices for a question to 2 possible answers, what is the chance of guessing the correct answer? _______________

3. How many of the 6 questions on page 272 do you think you answered correctly? ______ questions

4. Is it likely that you got all 6 correct? __________

5. Is it likely that you got all 6 incorrect? __________

6. Suppose each correct answer is worth 1 point and each incorrect answer carries a penalty of $\frac{1}{3}$ point. Complete the Total Points column of the table. You will complete the Class Tally column later.

Number Correct	Number Incorrect	Total Points	Class Tally
6	0	6	
5	1		
4	2		
3	3		
2	4	$2 - \frac{4}{3} = \frac{2}{3}$	
1	5		
0	6		

Date Time

LESSON 7·8 Guessing on Multiple-Choice Tests *continued*

7. For each question on the following test, first draw a line through each answer that you know is not correct. Then circle one answer for each question. If you do not know the correct answer, guess.

Number correct ______________

1. The neck of a 152-pound person weighs about

   A. 100 pounds.

   B. $12\frac{1}{2}$ pounds.

   C. $11\frac{1}{2}$ pounds.

   D. $10\frac{1}{2}$ pounds.

2. The average height of a full-grown weeping willlow tree is

   A. 50 feet.

   B. 45 feet.

   C. 35 feet.

   D. 2 feet.

3. The normal daily high temperature for July in Cleveland, Ohio, is

   A. 84°F.

   B. 82°F.

   C. 80°F.

   D. 0°F.

4. The circumference of Earth at the equator is about

   A. 24,901.6 miles.

   B. 24,801.6 miles.

   C. 24,701.6 miles.

   D. 2,000 miles.

5. In 2001, the average U.S. resident consumed about 138 pounds of which food?

   A. sugar

   B. spinach

   C. potatoes

   D. rice

6. A slice of white bread has about how many calories?

   A. 3

   B. 65

   C. 70

   D. 75

Date Time

LESSON 7•8

Guessing on Multiple-Choice Tests *continued*

8. When you can narrow the choices for a question to 3 possible answers, what is the chance of guessing the correct answer? ________

9. How many of the 6 questions do you think you answered correctly?

   _____ questions

10. Is it likely that you got all 6 correct? ________

11. Is it likely that you got all 6 incorrect? ________

12. Suppose each correct answer is worth 1 point and each incorrect answer carries a penalty of $\frac{1}{3}$ point. Complete the Total Points column of the table below.

Number Correct	Number Incorrect	Total Points	Class Tally
6	0	6	
5	1		
4	2		
3	3		
2	4	$2 - \frac{4}{3} = \frac{2}{3}$	
1	5		
0	6		

Date Time

LESSON 7·8

Algebraic Expressions

1. Write an algebraic expression for each situation. Use the suggested variable.

   a. Dakota is 14 inches taller than Gerry. If Gerry is h inches tall, how many inches tall is Dakota? ________ (unit)

   b. Sam ran for $\frac{4}{5}$ the length of time that Justin ran. If Justin ran r minutes, how long did Sam run? ________ (unit)

   c. Beyonce has x CDs in her collection. If Ann has 9 fewer CDs than Beyonce, how many CDs does Ann have? ________ (unit)

   d. Charlie has d dollars. Leanna has 6 times as much money as Charlie. How much money does Leanna have? ________ (unit)

   e. Erica has been a lifeguard for y years. That is 3 times as many years as Tom. How long has Tom worked as a lifeguard? ________ (unit)

2. Write the rule for the numbers in the table.

x	y
1.5	6.75
2	9
−3	−13.5
0.25	1.125

Rule ________

3. Write the rule for the numbers in the table.

a	b
$\frac{2}{3}$	$\frac{4}{15}$
$\frac{1}{6}$	$\frac{1}{15}$
$\frac{3}{4}$	$\frac{3}{10}$
$\frac{4}{5}$	$\frac{8}{25}$

Rule ________

Translate each situation from words into an algebraic expression. Then solve the problem that follows.

4. Calida has 7 more crayons than 3 times the number of crayons Royce has. If Royce has c crayons, how many does Calida have? ________

   If Royce has 12 crayons, how many does Calida have? ________ (unit)

5. Alinda has seen 4 fewer than $\frac{1}{2}$ the number of movies that her sister has seen. If her sister has seen m movies, how many has

   Alinda seen? ________

   If her sister has seen 20 movies, how many has Alinda seen? ________ (unit)

Date Time

LESSON 7·9

Math Boxes

1. Solve each equation.

Solution

a. $\frac{28}{c} = 14$ ____________

b. $\frac{15}{33} = \frac{5}{x}$ ____________

c. $\frac{500}{10,000} = \frac{p}{100}$ ____________

d. $\frac{25}{75} = \frac{d}{12}$ ____________

2. Complete.

a. 20% of 80 = ________

b. 75% of 48 = ________

c. 55% of 1,000 = ________

d. 30% of 250 = ________

3. Use the graph to answer the following questions.

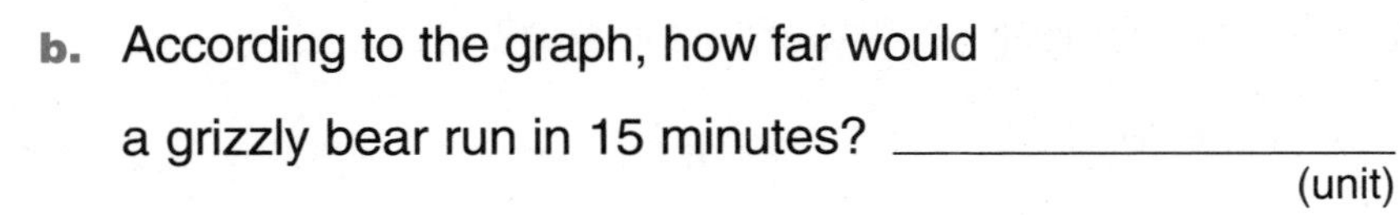

a. Suppose a black rhino runs at its top speed for 6 minutes.

How far would the rhino travel? ____________ (unit)

b. According to the graph, how far would a grizzly bear run in 15 minutes? ____________ (unit)

c. Which animal's speed is about 20% the speed of the cheetah? ____________

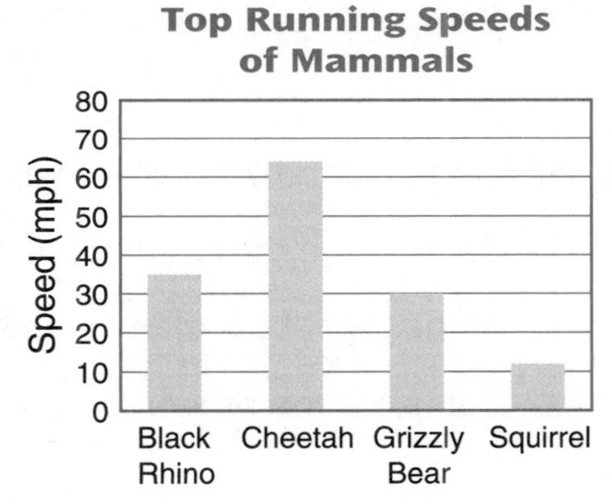

4. Divide. Express the remainder as a fraction in simplest form.

$75\overline{)7,698}$

7,698 ÷ 75 = ____________

5. Express each ratio as a fraction in simplest form.

a. 100 to 60 ________

b. 80 out of 56 ________

c. 1,440 calories in 8 ounces ________

d. 112 males for every 98 females ________

Date _______________ Time _______________

Solving Rate Problems

Math Message

1. A computer printer prints 70 pages in 2 minutes. How many pages will it print in 5 minutes? _______________

2. Anevay trains at an indoor track during the winter. She can run 24 laps in 8 minutes. At this rate, how many laps can Anevay run in 12 minutes? _______________

---

3. Ms. Marquez is reading stories that her students wrote. She has read 5 stories in 40 minutes.

   a. At this rate, how long would it take her to read 1 story? _______________

   b. How long will it take her to read all 30 of her students' stories? _______________

   c. Complete the proportion to show your solution. $\frac{5 \text{ stories}}{40 \text{ minutes}} = \frac{30 \text{ stories}}{\square \text{ minutes}}$

4. Roshaun scored 75 points in the first 5 basketball games.

   a. On average, how many points did he score per game? _______________

   b. At this rate, how many points might he score in a 15-game season? _______________

   c. Complete the proportion to show your solution. $\frac{75 \text{ points}}{5 \text{ games}} = \frac{\square \text{ points}}{15 \text{ games}}$

5. Last year, 55 students sold $1,210 worth of candy for their band's fund-raiser.

   a. On average, how many dollars' worth of candy did each student sell? _______________

   b. This year, 67 students will be selling candy. If they sell at the same rate as last year, how much money can they expect to raise? _______________

   c. Complete the proportion to show your solution. $\frac{\square \text{ students}}{\$\square} = \frac{\square \text{ students}}{\$\square}$

6. Anoki worked at the checkout counter from 5:30 P.M. to 11 P.M. He earned $33.

   a. How much did he earn per hour? _______________

   b. Anoki works $27\frac{1}{2}$ hours per week. How much will he earn in 1 week? _______________

   c. Complete the proportion to show your solution. $\frac{\$\square}{\square \text{ hours}} = \frac{\$\square}{\square \text{ hours}}$

Date Time

LESSON 8•1

Solving Rate Problems *continued*

7. The furlong is a unit of distance commonly used in horse racing. There are 40 furlongs in 5 miles.

a. Fill in the rate table.

miles	1	2		5	8	10
furlongs			24	40		

b. How many furlongs are in 8 miles? ______________

Complete the proportion to show your solution.

$\frac{\square \text{ miles}}{\square \text{ furlongs}} = \frac{\square \text{ miles}}{\square \text{ furlongs}}$

c. How many miles are in 8 furlongs?

____

Complete the proportion to show your solution.

$\frac{\square \text{ miles}}{\square \text{ furlongs}} = \frac{\square \text{ miles}}{\square \text{ furlongs}}$

8. Nico's grandfather read a 240-page book in 6 hours.

a. Fill in the rate table.

pages		120		240	80	
hours	1		4	6		8

b. At this rate, how long would it take him to read a 160-page book? ______________

Complete the proportion to show your solution.

$\frac{\square \text{ pages}}{\square \text{ hours}} = \frac{\square \text{ pages}}{\square \text{ hours}}$

c. How many hours did it take him to read 80 pages? ______________

Complete the proportion to show your solution.

$\frac{\square \text{ pages}}{\square \text{ hours}} = \frac{\square \text{ pages}}{\square \text{ hours}}$

Use any method you wish to solve the following problems. Write a proportion to show your solution.

9. A recipe for a 2-pound loaf of bread calls for 4 cups of flour. How many 2-pound loaves can you make with 12 cups of flour? ______________

$\frac{\square \text{ loaves}}{\square \text{ cups}} = \frac{\square \text{ loaves}}{\square \text{ cups}}$

10. Two inches of rain fell between 7 A.M. and 3 P.M. The rain continued at the same rate until 7 P.M.

How many inches of rain fell between 7 A.M. and 7 P.M.? ______________

$\frac{\square \text{ inches}}{\square \text{ hours}} = \frac{\square \text{ inches}}{\square \text{ hours}}$

Date Time

LESSON 8•1

More Division Practice

Divide. Show your work in the space below.
For Problems 1–3, round answers to the nearest tenth.

1. 571 ÷ 8 = ________
2. 2,723 / 94 = ________
3. 815 ÷ 46 = ________

For Problems 4–6, round answers to the nearest hundredth.

4. 89 / 6 = ________
5. 3,714 / 42 = ________
6. 217 ÷ 18 = ________

7. Write a number story for Problem 6.

____

____

____

____

____

____

Date Time

LESSON 8•1

Math Boxes

1. Express each rate as a per-unit rate.

   a. 108 words in 4 minutes

   $\frac{\square \text{ words}}{1 \text{ minute}}$

   b. 300 miles in 15 hours

   ☐ miles/hour

   c. $102.50 for 10 hours

   $ ☐ /hour

2. Tamika earns $500 for 40 hours of work.

   Fill in the rate table.

hours	5	10	15
amount ($)	62.50		

   At this rate, how much will Tamika earn for 55 hours of work?

   ____

3. For lunch, the school cafeteria offers a main course and a beverage. For the main course, students can choose spaghetti, hamburgers, or hot dogs. The beverage choices are milk, soda, and juice. Draw a tree diagram to show all the possible meal combinations.

   If you choose a meal at random, what is the probability of getting a

   a. hot dog?

   ____

   b. hot dog and juice? ________

4. Add or subtract.

   a. $14 + (-72) =$ ________

   b. ________ $= 27 - (-28)$

   c. ________ $= -63 + (-87)$

   d. $-33 - (-89) =$ ________

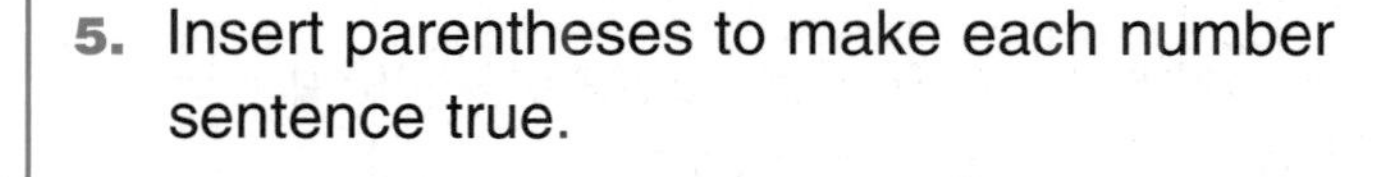

5. Insert parentheses to make each number sentence true.

   a. $5 * 10^2 + 10^2 * 2 = 2{,}000$

   b. $9 * 3 + 4 = 63$

   c. $16 + 2^2 - 5 + 3 = 12$

   d. $7 * 4 - 2 + 1 = 21$

Date Time

LESSON 8•2

Rate Problems and Proportions

Math Message

Solve the problems below. Then write a proportion for each problem.

1. Robin rode her bike at an average speed of 8 miles per hour. At this rate, how far would she travel in 3 hours? ____________

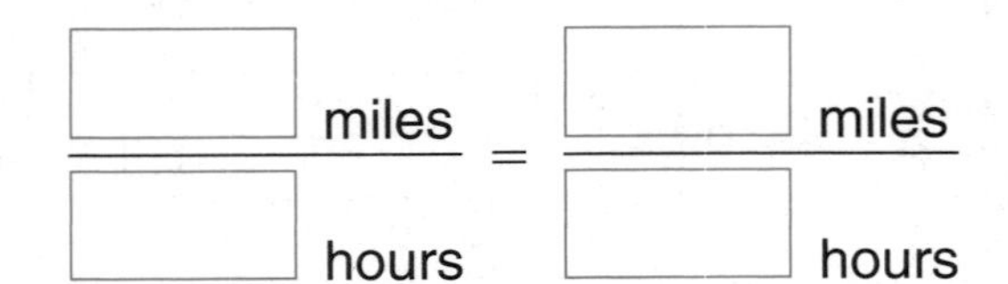

2. A high-speed copier makes 90 copies per minute. How long will it take to make 270 copies? ____________

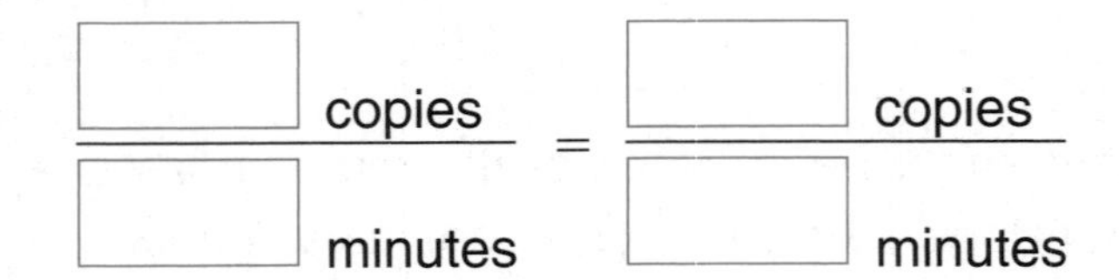

3. Talia rode her bike at an average speed of 10 miles per hour. At this rate, how long would it take Talia to ride 15 miles? ____________

$\frac{\square \text{ miles}}{\square \text{ hours}} = \frac{\square \text{ miles}}{\square \text{ hours}}$

For each of the following problems, first complete the rate table. Use the table to write an open proportion. Solve the proportion. Then write the answer to the problem. Study the first problem.

4. Angela earns \$6 per hour babysitting. How long must she work to earn \$72?

   Answer: Angela must work

   ________ hours to earn \$72.

dollars	6	72
hours	1	t

$\frac{6 * 12}{1 * 12} = \frac{72}{t}$

5. There are 9 calories per gram of fat. How many grams of fat have 63 calories?

   Answer: ________ grams of fat have 63 calories.

calories		
grams (of fat)		

$\frac{\square}{\square} = \frac{\square}{\square}$

Date Time

Rate Problems and Proportions *continued*

6. If carpet costs $22.95 per square yard, how much will 12 square yards of carpet cost?

dollars		
square yards		

Answer: 12 square yards of carpet will cost __________.

7. There are 80 calories in 1 serving of soup. How many servings of soup have 120 calories?

calories		
servings		

Answer: __________ servings of soup have 120 calories.

8. A car goes 480 miles on a 12-gallon tank of gas. How many miles is this per gallon?

miles		
gallons		

Answer: The car will go __________ miles on 1 gallon of gas.

9. Liang read the first 48 pages of a mystery novel in 3 hours. At this rate, how long will it take him to read 80 pages?

hours		
pages		

Answer: It will take Liang __________ hours to read 80 pages.

10. A TV station runs 42 minutes of commercials in 7 half-hour programs. How many minutes of commercials does it run per hour?

commercial minutes		
hours		

Answer: The station runs __________ minutes of commercials per hour.

Date Time

LESSON 8·2

Dividing Decimals by Decimals: Part 2

In general, you can think of a fraction as a division problem.

The fraction $\frac{3}{8}$ equals $3 \div 8$ or $8\overline{)3}$.

$\frac{3}{8}$ ← numerator → $8\overline{)3}$
← denominator →

As you do with a fraction, you can multiply the dividend and divisor by the same number without changing the value of the quotient.

$$\frac{17.48}{0.076} * \frac{1,000}{1,000} = \frac{17,480}{76}$$

original problem — equivalent problem

$0.076\overline{)17.480}$ $76\overline{)17,480}$

Rename each division problem to make the divisor a whole number. Then solve the equivalent problem using the method of your choice.

Equivalent Problem

1. $3.5\overline{)27.3}$ = ____________________

Quotient = __________

Equivalent Problem

2. $1.9\overline{)7.03}$ = ____________________

Quotient = __________

Equivalent Problem

3. $5.06\overline{)47.058}$ = ____________________

Quotient = __________

Equivalent Problem

4. $0.56\overline{)21.28}$ = ____________________

Quotient = __________

Date Time

LESSON 8•2

Math Boxes

1. The formula $d = r * t$ gives the distance d traveled at speed r in time t. Use this formula to solve the problem below.

The distance from San Francisco to Los Angeles is about 420 miles. About how many hours would it take to drive from San Francisco to Los Angeles at a constant rate of 55 miles per hour? Round your answer to the nearest tenth.

$t =$ ________

2. Solve.

Solution

a. $\frac{42}{18} = \frac{x}{3}$ ________

b. $\frac{125}{n} = \frac{5}{1}$ ________

c. $\frac{36}{w} = \frac{18}{2}$ ________

d. $\frac{k}{150} = \frac{3}{5}$ ________

e. $\frac{90}{3} = \frac{9}{u}$ ________

3. Thirty-three sixth graders at Maple Middle School belong to the drama club. Thirty-seven sixth graders are in the school choir.

Use this information to complete the Venn diagram below.

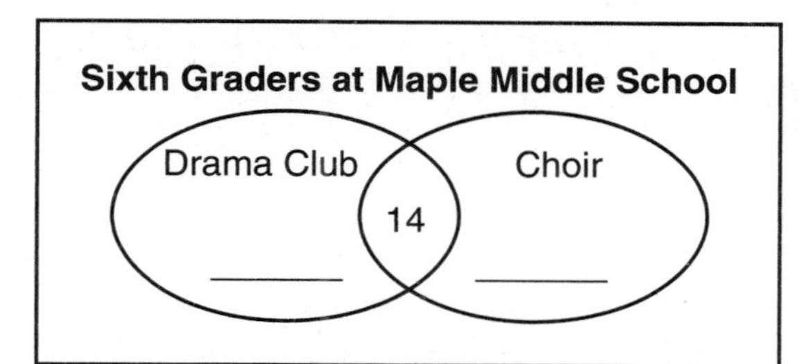

4. Solve.

a. $\frac{-9 * 5}{3} =$ ________

b. ________ $= (-1)(-7)^2$

c. $\frac{-6 + (-3) + (-7)}{4} =$ ________

5. Fill in the blanks.

a. $-8 +$ ________ $= 0$

b. $\frac{2}{3} +$ ________ $= 0$

c. $-5 -$ ________ $= 0$

Date Time

LESSON 8·3

Equivalent Fractions and Cross Products

Math Message

SRB 114 115

For Part a of each problem, write = or ≠ in the answer box.
For Part b, calculate the cross products.

1. a. $\frac{3}{5}$ ☐ $\frac{6}{10}$

b. $10 * 3 =$ ______ ______ $= 5 * 6$

$\frac{3}{5} \times \frac{6}{10}$

2. a. $\frac{7}{8}$ ☐ $\frac{2}{3}$

b. $3 * 7 =$ ______ ______ $= 8 * 2$

$\frac{7}{8} \times \frac{2}{3}$

3. a. $\frac{2}{3}$ ☐ $\frac{6}{9}$

b. $9 * 2 =$ ______ ______ $= 3 * 6$

$\frac{2}{3} \times \frac{6}{9}$

4. a. $\frac{6}{9}$ ☐ $\frac{8}{12}$

b. $12 * 6 =$ ______ ______ $= 9 * 8$

$\frac{6}{9} \times \frac{8}{12}$

5. a. $\frac{2}{8}$ ☐ $\frac{4}{10}$

b. $10 * 2 =$ ______ ______ $= 8 * 4$

$\frac{2}{8} \times \frac{4}{10}$

6. a. $\frac{10}{12}$ ☐ $\frac{5}{8}$

b. $8 * 10 =$ ______ ______ $= 12 * 5$

$\frac{10}{12} \times \frac{5}{8}$

7. a. $\frac{1}{4}$ ☐ $\frac{5}{20}$

b. $20 * 1 =$ ______ ______ $= 4 * 5$

$\frac{1}{4} \times \frac{5}{20}$

8. a. $\frac{5}{7}$ ☐ $\frac{15}{21}$

b. $21 * 5 =$ ______ ______ $= 7 * 15$

$\frac{5}{7} \times \frac{15}{21}$

9. a. $\frac{10}{16}$ ☐ $\frac{4}{8}$

b. $8 * 10 =$ ______ ______ $= 16 * 4$

$\frac{10}{16} \times \frac{4}{8}$

10. a. $\frac{3}{5}$ ☐ $\frac{10}{15}$

b. $15 * 3 =$ ______ ______ $= 5 * 10$

$\frac{3}{5} \times \frac{10}{15}$

11. What pattern can you find in Parts a and b in the problems above?

____

Date Time

LESSON 8•3

Math Boxes

1. Which rate is equivalent to 70 km in 2 hr 30 min? Fill in the circle next to the best answer.

   ○ **A.** 35 km in 75 min

   ○ **B.** 70,000 m in 230 min

   ○ **C.** 140 km in 4 hr 30 min

   ○ **D.** 1,400 m in 300 min

2. A boat traveled 128 kilometers in 4 hours.

   Fill in the rate table.

distance (km)	24		72		144
hours	$\frac{3}{4}$	$1\frac{1}{2}$		3	

   At this rate, how far did the boat travel in 2 hours 15 minutes?

   ____

3. A bag contains 1 red counter, 2 blue counters, and 1 white counter. You pick 1 counter at random. Then you pick a second counter without replacing the first counter.

   a. Draw a tree diagram to show all possible counter combinations.

   b. What is the probability of picking 1 red counter and 1 white counter (in either order)? ________

4. Add or subtract.

   a. $-303 + (-28) =$ __________

   b. __________ $= 245 - 518$

   c. __________ $= -73 + 89$

   d. $280 - (-31) =$ __________

5. Insert parentheses to make each number sentence true.

   a. $0.01 * 7 + 9 / 4 = 0.04$

   b. $\frac{4}{5} * 25 - 10 / 2 = 15$

   c. $\sqrt{64} / 5 + 3 * 3 = 3$

   d. $5 * 10^2 + 10^2 * 2 = 700$

Date Time

Solving Proportions with Cross Products

Use cross multiplication to solve these proportions.

SRB 114 115

Example: $\frac{4}{6} = \frac{p}{15}$

$15 * 4 =$ ___ ___ $= 6 * p$

$\frac{4}{6} = \frac{p}{15}$

$15 * 4 = 6 * p$

$60 = 6p$

$\frac{60}{6} = p$

$10 = p$

1. $\frac{3}{6} = \frac{y}{10}$ ______________

2. $\frac{7}{21} = \frac{3}{c}$ ______________

3. $\frac{m}{20} = \frac{2}{8}$ ______________

4. $\frac{2}{10} = \frac{5}{z}$ ______________

5. $\frac{9}{15} = \frac{12}{k}$ ______________

6. $\frac{10}{12} = \frac{d}{9}$ ______________

For each problem on the next page, set up a proportion and solve it using cross multiplication. Then write the answer.

Example: Jessie swam 6 lengths of the pool in 4 minutes. At this rate, how many lengths will she swim in 10 minutes?

Proportion: $\frac{\text{6 lengths}}{\text{4 minutes}} = \frac{n \text{ lengths}}{\text{10 minutes}}$

Solution: $\frac{6}{4} = \frac{n}{10}$

$10 * 6 =$ ___ ___ $= 4 * n$

$\frac{6}{4} = \frac{n}{10}$

$10 * 6 = 4 * n$

$60 = 4n$

$\frac{60}{4} = n$

$15 = n$

Answer: Jessie will swim __________ lengths in 10 minutes.

Date Time

LESSON 8·3

Solving Proportions with Cross Products *continued*

SRB 114 115

7. Belle bought 8 yards of ribbon for $6.
How many yards could she buy for $9?

Solution:

$\frac{\square}{\square} = \frac{\square}{\square}$

Answer: Belle could buy ________ yards of ribbon for $9.

8. Before going to France, Maurice exchanged $25 for 20 euros. At that exchange rate, how many euros could he get for $80?

Solution:

$\frac{\square}{\square} = \frac{\square}{\square}$

Answer: Maurice could get ________ euros for $80.

9. One gloomy day, 4 inches of rain fell in 6 hours. At this rate, how many inches of rain had fallen after 4 hours?

Solution:

$\frac{\square}{\square} = \frac{\square}{\square}$

Answer: ________ inches of rain had fallen in 4 hours.

10. Adelio's apartment building has 9 flights of stairs. To climb to the top floor, he must go up 144 steps. How many steps must he climb to get to the fifth floor?

Solution:

$\frac{\square}{\square} = \frac{\square}{\square}$

Answer: Adelio must climb ________ steps.

11. At sea level, sound travels 0.62 mile in 3 seconds. What is the speed of sound in miles per hour? (*Hint:* First find the number of seconds in 1 hour.)

Solution:

$\frac{\square}{\square} = \frac{\square}{\square}$

Answer: Sound travels at the rate of ________ miles per hour.

Date _______________ Time _______________

Rate * Time = Distance

Math Message

For each problem, make a rate table. Then write a number model and solve it.

1. Grandma Riley drove her car at 60 miles per hour for 4 hours. How far did she travel?

   Number model _______________

   Answer: She traveled _______________ miles in 4 hours.

2. A bamboo plant grows 8 inches per day. How tall will it be after 7 days?

   Number model _______________

   Answer: The plant will be _______________ inches tall.

3. A rocket is traveling at 40,000 miles per hour. How far will it travel in 168 hours?

   Number model _______________

   Answer: The rocket will travel _______________ miles in 168 hours.

4. Amora can ride her bicycle at 9 miles per hour. At this rate, how long will it take her to ride 30 miles?

   Number model _______________

   Answer: It will take her _______________ hours to ride 30 miles.

5. Australia is moving about 3 inches per year with respect to the southern Pacific Ocean. How many *feet* will it move in 50 years?

   Number model _______________

   Answer: Australia will move _______________ feet in 50 years.

Date Time

LESSON 8•4

Using Unit Fractions to Find a Whole

Example 1:

Alex collects sports cards. Seventy of the cards feature basketball players. These 70 cards are $\frac{2}{3}$ Alex's collection. How many sports cards does Alex have?

- If $\frac{2}{3}$ the collection is 70 cards, then $\frac{1}{3}$ is 35 cards.
- Alex has all the cards—that's $\frac{3}{3}$ the cards.
- Therefore, Alex has 3 $*$ 35, or 105 cards.

Example 2:

Barb's grandmother baked cookies. She gave Barb 12 cookies, which was $\frac{2}{5}$ the total number she baked. How many cookies did Barb's grandmother bake?

- If $\frac{2}{5}$ the total is 12 cookies, then $\frac{1}{5}$ is 6 cookies.
- Barb's grandmother baked all the cookies—that's $\frac{5}{5}$ the cookies.
- She baked 5 $*$ 6, or 30 cookies.

1. Six jars are filled with cookies. The number of cookies in each jar is not known. For each clue given in the table, find the number of cookies in the jar.

Clue	Number of Cookies in Jar
$\frac{1}{2}$ jar contains 31 cookies.	
$\frac{2}{8}$ jar contains 10 cookies.	
$\frac{3}{5}$ jar contains 36 cookies.	
$\frac{3}{8}$ jar contains 21 cookies.	
$\frac{4}{7}$ jar contains 64 cookies.	
$\frac{3}{11}$ jar contains 45 cookies.	

2. Jin is walking to a friend's house. He has gone $\frac{6}{10}$ the distance in 48 minutes. If he continues at the same speed, about how long will the entire walk take?

____

3. A candle burned $\frac{3}{8}$ the way down in 36 minutes. If it continues to burn at the same rate, about how many more minutes will the candle burn before it is used up?

____

Date Time

LESSON 8·4 How Many Calories Do You Use Per Day?

Your body needs food. It uses the materials in food to produce energy—energy to keep your body warm and moving, to live and grow, and to build and repair muscles and tissues.

The amount of energy a food will produce when it is digested by the body is measured in a unit called the **calorie.** A calorie is not a substance in food.

1. The following table shows the number of calories used per minute and per hour by the average sixth grader for various activities. Complete the table. Round your answers for calories per minute to the nearest tenth and calories per hour to the nearest ten.

Calorie Use by Average Sixth Graders

Activity	Calories/Minute (to nearest 0.1)	Calories/Hour (to nearest 10)
Sleeping	0.7	40
Studying, writing, sitting	1.2	70
Eating, talking, sitting in class	1.2	70
Standing	1.3	80
Dressing, undressing		90
Watching TV	1.0	60
Walking (briskly, at 3.5 mph)	3.0	180
Doing housework, gardening	2.0	
Playing the piano	2.7	160
Raking leaves	3.7	220
Shoveling snow	5.0	300
Bicycling (6 mph)		170
Bicycling (13 mph)	4.5	
Bicycling (20 mph)	8.3	
Running (5 mph)	6.0	360
Running (7.5 mph)		560
Swimming (20 yd/min)	3.3	200
Swimming (40 yd/min)	5.8	350
Basketball, soccer (vigorous)	9.7	580
Volleyball	4.0	240
Aerobic dancing (vigorous)	6.0	360
Bowling	3.4	200

Date Time

LESSON 8•4

How Many Calories Do You Use Per Day? *continued*

2. Think of all the things you do during a typical 24-hour day during which you go to school.

a. List your activities in the table below.

b. Record your estimate of the time you spend on each activity (to the nearest 15 minutes). Be sure the times add up to 24 hours.

c. For each activity, record the number of calories used per minute or per hour. Then calculate the number of calories you use for the activity.

Example:

Suppose you spend 8 hours and 15 minutes sleeping.
Choose the per-hour rate: Sleeping uses 40 calories per hour.
Multiply: 8.25 hours ∗ 40 calories per hour = 330 calories

My Activities during a Typical School Day (24 hr)

Activity	Time Spent on Activity	Calorie Rate (cal/min or cal/hr)	Calories Used for Activity

3. After you complete the table, find the total number of calories you use in 24 hours.

In a typical 24-hour day during which I go to school, I use about ____________ calories.

Date Time

LESSON 8•4

Math Boxes

1. The formula $d = r * t$ gives the distance d traveled at speed r in time t. Use this formula to solve the problem below.

Which formula is equivalent to $d = r * t$? Choose the best answer.

- ⬭ $r = \frac{d}{t}$
- ⬭ $r = d - t$
- ⬭ $r = \frac{t}{d}$
- ⬭ $r = t + d$

2. Solve.

Solution

a. $\frac{6}{p} = \frac{2}{7}$ ____________

b. $\frac{f}{2} = \frac{3}{12}$ ____________

c. $\frac{4}{15} = \frac{12}{x}$ ____________

d. $\frac{24}{36} = \frac{6}{w}$ ____________

e. $\frac{7}{8} = \frac{y}{2}$ ____________

3. Of the 330 students at Pascal Junior High, 45 run track and 67 play basketball. Twenty-two students participate in both sports.

Use this information to complete the Venn diagram below. Label each set (ring). Write the number of students belonging to each individual set and the intersection of the sets.

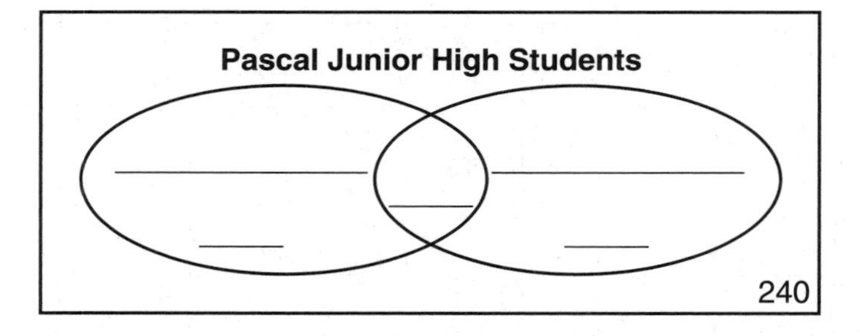

4. Solve.

a. $-\frac{32}{2} + \frac{-75}{-15} =$ ____________

b. $(-9)^2 (-1)^5 =$ ____________

c. $\frac{-68 - 112}{-10} =$ ____________

5. Fill in the blanks. (*Hint*: For decimals, think fractions.)

a. $\frac{7}{3} *$ ____________ $= 1$

b. $0.01 *$ ____________ $= 1$

c. $0.5 *$ ____________ $= 1$

Date Time

LESSON 8•5 Food Nutrition Labels

Use the information from the food label at the right to complete the following statements.

Low-fat yogurt

Nutrition Facts	
Serving Size 1 container (227 g)	
Amount Per Serving	
Calories 240 Calories from Fat 27	
	% Daily Value
Total Fat 3 g	**5%**
Saturated Fat 1.5 g	**8%**
Cholesterol 15 mg	**5%**
Sodium 150 mg	**6%**
Potassium 450 mg	**13%**
Total Carbohydrate 44 g	**15%**
Dietary Fiber 1 g	**4%**
Sugars 43 g	
Protein 9 g	
Vitamin A 2% • Vitamin C 10%	
Calcium 35% • Iron 0%	
Calories per gram: Fat 9 • Carbohydrate 4 • Protein 4	

1. There are 240 calories per serving. Of these 240 calories,

   ________ calories come from fat.

2. There are ________ grams of total carbohydrate per serving.

   Complete the proportion.

   $$\frac{\text{1 g of carbohydrate}}{\text{4 calories}} = \frac{\square\text{ g of carbohydrate}}{\square\text{ calories}}$$

   Solve the proportion. How many calories come from carbohydrate?

   ________ calories

3. Write the unit rate you can use to calculate calories from protein.

   ____

For each food label below, record the number of calories from fat. Then calculate the numbers of calories from carbohydrate and from protein. Add to find the total calories per serving.

White bread

Nutrition Facts	
Serving Size 1 slice (23 g)	
Servings Per Container 20	
Amount Per Serving	
Calories 65 Calories from Fat 9	
	% Daily Value
Total Fat 1 g	**2%**
Total Carbohydrate 12 g	**4%**
Protein 2 g	

Hot dog

Nutrition Facts	
Serving Size 1 link (45 g)	
Servings Per Container 10	
Amount Per Serving	
Calories 150 Calories from Fat 120	
	% Daily Value
Total Fat 13 g	**20%**
Total Carbohydrate 1 g	**<1%**
Protein 7 g	

4. Calories

   From fat ________

   From carbohydrate ________

   From protein ________

   Total calories ________

5. Calories

   From fat ________

   From carbohydrate ________

   From protein ________

   Total calories ________

Date Time

LESSON 8•5 Plan Your Lunch

1. Choose 5 items you would like to have for lunch from the following menu. Choose your favorite foods—pay no attention to calories. Make a check mark next to each item.

Food	Total Calories	Calories from Fat	Calories from Carbohydrate	Calories from Protein
Ham sandwich	265	110	110	45
Turkey sandwich	325	70	155	100
Hamburger	330	135	120	75
Cheeseburger	400	200	110	90
Double burger, cheese, sauce	500	225	175	100
Grilled cheese sandwich	380	220	100	60
Peanut butter and jelly sandwich	380	160	170	50
Chicken nuggets (6)	250	125	65	60
Bagel	165	20	120	25
Bagel with cream cheese	265	105	125	35
Hard-boiled egg	80	55	0	25
French fries (small bag)	250	120	115	15
Apple	100	10	90	0
Carrot	30	0	25	5
Orange	75	0	70	5
Cake (slice)	235	65	160	10
Cashews (1 oz)	165	115	30	20
Doughnut	200	100	75	25
Blueberry muffin	110	30	70	10
Apple pie (slice)	250	125	115	10
Frozen-yogurt cone	100	10	75	15
Orange juice (8 fl oz)	110	0	104	8
2% milk (8 fl oz)	145	45	60	40
Skim milk (8 fl oz)	85	0	50	35
Soft drink (8 fl oz)	140	0	140	0
Diet soft drink (8 fl oz)	0	0	0	0

Date Time

LESSON 8•5

Plan Your Lunch *continued*

2. In the table below, record the 5 items you chose. Fill in the rest of the table and write the total number of calories for each column.

Food	Total Calories	Calories from Fat	Calories from Carbohydrate	Calories from Protein
Total				

What percent of the total number of calories in your lunch comes from fat? ________

From carbohydrate? ________ From protein? ________

3. Suppose a nutritionist recommends that, at most, 20–35% of the total number of calories should come from fat, about 35% from protein, and no more than 45–65% from carbohydrate.

Does the lunch you chose meet these recommendations? ________

4. Plan another lunch. This time use only 4 foods and try to limit the percent of calories from fat to between 20 and 35%, from protein to between 25 and 36%, and from carbohydrate to between 45 and 65%.

Food	Total Calories	Calories from Fat	Calories from Carbohydrate	Calories from Protein
Total				

What percent of the total number of calories in your lunch comes from fat? ________

From carbohydrate? ________ From protein? ________

LESSON 8•5

Unit Percents to Find a Whole

Example 1:

The sale price of a CD player is $120, which is 60% of its list price.
What is the list price?

- If 60% of the list price is $120, then 1% is $2. (120 / 60 = 2)
- The list price (the whole, or 100%) is $200. (100 * 2 = 200)

Example 2:

A toaster is on sale for $40, which is 80% of its list price.
What is the list price?

- If 80% of the list price is $40, then 1% is $0.50. (40 / 80 = 0.5)
- The list price (100%) is $50. (100 * 0.5 = 50)

Use your percent sense to estimate the list price for each item below.
Then calculate the list price. (*Hint:* First use your calculator to find what 1% is worth.)

Sale Price	Percent of List Price	Estimated List Price	Calculated List Price
$120	60%	$180	$200
$100	50%		
$8	32%		
$255	85%		
$77	55%		
$80	40%		
$9	60%		
$112.50	75%		
$450	90%		

Date Time

LESSON 8•5

Math Boxes

1. If there are 975 calories in 3 oz of ice cream, how many calories are in 4 oz of ice cream?

Set up a proportion below.

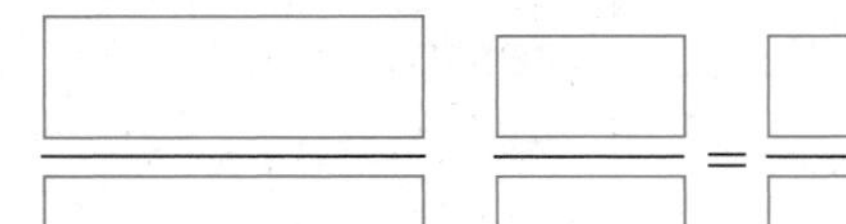

Solve the proportion in the space below.
Solution:

There are ____________ calories in 4 oz of ice cream.

2. Solve.

Solution

a. $\frac{c}{15} = \frac{21}{9}$ ________________

b. $\frac{15}{20} = \frac{18}{m}$ ________________

c. $\frac{w}{15} = \frac{14}{10}$ ________________

3. Use order of operations to evaluate each expression.

a. $9 * 5 / 10 + 3 - 2 =$ ________

b. $8 - 6 * 4 + 8 / 2 =$ ________

c. ________ $= 2 + 2 * 12 + 3^2 - 5$

d. ________ $= 15 / (2 + 3) - 8 * 2$

e. ________ $= 5^2 * 2 + 9 * 2$

4. Which equation describes the relationship between the numbers in the table? Circle the best answer.

x	y
0.55	$\frac{1}{2}$
0.6	1
1	5
1.5	10

A. $(y + 0.1) * \frac{1}{2} = x$

B. $(y * 0.1) + \frac{1}{2} = x$

C. $\frac{0.1}{2}y = x$

D. $(y + \frac{1}{2}) * 0.1 = x$

5. The area of square *CAMP* is 25 cm². Squares *CAMP* and *DAME* are congruent.

What is the area of triangle *PAE?*

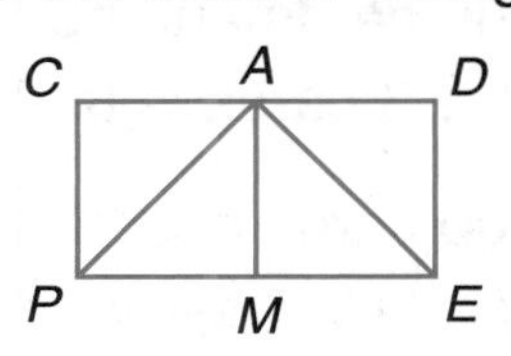

Area = ________ cm^2

Date Time

LESSON 8·6 Ratios

Math Message

Work with a partner. You may use a deck of cards to help you with these problems.

1. There are 2 facedown cards for every faceup card.
   If 6 of the cards are faceup, how many cards are facedown? ________ cards

2. You have 12 cards. One out of every 4 cards is faceup.
   The rest are facedown. How many cards are faceup? ________ cards

3. There are 4 facedown cards for every 3 faceup cards.
   If 8 of the cards are facedown, how many cards are faceup? ________ cards

4. Three out of every 5 cards are faceup. If 12 cards are faceup, how many cards are there in all? ________ cards

5. There are 2 faceup cards for every 5 facedown. If there are 21 cards in all, how many cards are faceup? ________ cards

6. The table at the right shows the average number of wet days in selected cities for the month of October.

   a. How many more wet days does Moscow have than Beijing? ________

   b. Moscow has how many times as many wet days as Beijing? ________

   c. The number of wet days in Beijing is what fraction of the number of wet days in Sydney? ________

City	Wet Days
Beijing, China	3
Boston, United States	9
Frankfurt, Germany	14
Mexico City, Mexico	13
Moscow, Russia	15
Sydney, Australia	12

Try This

7. You have 5 faceup cards and no facedown cards. You add some facedown cards so 1 in every 3 cards is faceup. How many cards are there now? ________ cards

8. You have 5 faceup cards and 12 facedown cards. You add some faceup cards so 2 out of every 5 cards are faceup. How many cards are there now? ________ cards

9. You have 8 faceup cards and 12 facedown cards. You add some faceup cards so $\frac{2}{3}$ of the cards are faceup.

   How many cards are faceup? ________ cards Facedown? ________ cards

Date Time

LESSON 8•6 Math Boxes

1. Solve.

a. If 9 counters are $\frac{1}{6}$ of a set, how many counters are in the set?

______

b. If $\frac{5}{7}$ of a mystery number is 80, then $\frac{1}{7}$ of the mystery number is ______.

The mystery number is ______.

2. An artist mixes yellow paint with blue paint to make a certain shade of green. The ratio of yellow to blue is 3 to 5. How much yellow paint should the artist mix with 20 ounces of blue paint?

Write a proportion. Then solve.

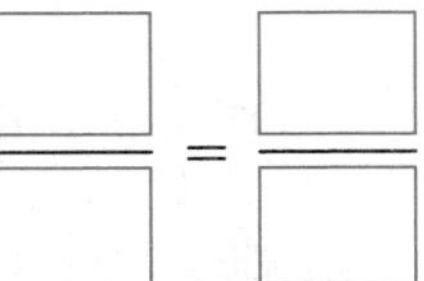

______ ounces of yellow paint

3. Janine watches about 12 hours of television per week. Complete the table. Then use your protractor to make a circle graph of the information.

Type of Show	Number of Hours	Percent of Hours	Degrees
Comedy	4		
Educational	1		
News	2		
Sports	3		
Cartoon	2		
Total			

______ (title)

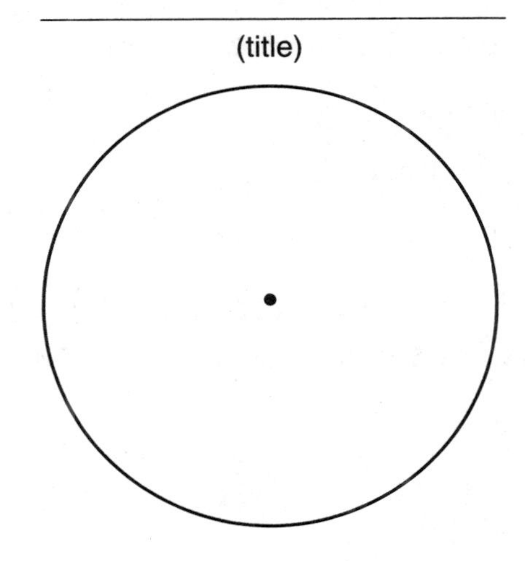

4. Use the general pattern

$a(b + c) = (a * b) + (a * c)$

to compute the following products mentally. One has been done for you.

$3 * 1\frac{2}{3} = (3 * 1) + (3 * \frac{2}{3}) = 3 + 2 = 5$

a. $9 * 2\frac{1}{9} =$ ______

b. $5 * 2\frac{3}{10} =$ ______

5. The formula $F = \frac{9}{5}C + 32$ can be used to convert temperatures from Celsius to Fahrenheit.

F is the temperature in degrees Fahrenheit.
C is the temperature in degrees Celsius.

Calculate the temperature in degrees Fahrenheit for the following Celsius temperatures.

a. 20°C = ______°F

b. −25°C = ______°F

Date Time

LESSON 8•6 Ratio Number Stories

For each problem, write and solve a proportion.

1. A spinner lands on blue 4 times for every 6 times it lands on green. How many times does it land on green if it lands on blue 12 times?

   a. $\frac{\text{(lands on blue)}}{\text{(lands on green)}}$ $\frac{\square}{\square} = \frac{\square}{\square}$

   b. It lands on green ______ times.

2. It rained 2 out of 5 days in the month of April. On how many days did it rain that month?

   a. $\frac{\text{(days with rain)}}{\text{(days)}}$ $\frac{\square}{\square} = \frac{\square}{\square}$

   b. It rained ______ days in April.

3. Of the 42 animals in the Children's Zoo, 3 out of 7 are mammals. How many mammals are in the Children's Zoo?

   a. $\frac{\text{(mammals)}}{\text{(animals)}}$ $\frac{\square}{\square} = \frac{\square}{\square}$

   b. Answer: ______________

4. Last week, Mikasi spent 2 hours doing homework for every 3 hours he watched TV. If he spent 6 hours doing homework, how many hours did he spend watching TV?

   a. $\frac{\square}{\square}$ $\frac{\square}{\square} = \frac{\square}{\square}$

   b. Answer: ______________

5. Five out of 8 students at Lane Junior High School play a musical instrument. If 140 students play an instrument, how many students attend Lane School?

   a. $\frac{\square}{\square}$ $\frac{\square}{\square} = \frac{\square}{\square}$

   b. Answer: ______________

6. A choir has 50 members. Twenty members are sopranos. How many sopranos are there for every 5 members of the choir?

   a. $\frac{\square}{\square}$ $\frac{\square}{\square} = \frac{\square}{\square}$

   b. Answer: ______________

Date Time

LESSON 8•6

Ratio Number Stories *continued*

7. Mr. Dexter sells subscriptions to a magazine for $18 each. For each subscription he sells, he earns $8. One month, he sold $900 in subscriptions. How much did he earn?

a. $\frac{\square}{\square}$ $\frac{\square}{\square} = \frac{\square}{\square}$

b. Answer: ______

8. At Kozminski School, the ratio of weeks of school to weeks of vacation is 9 to 4. How many weeks of vacation do students at the school get in 1 year?

a. Complete the table.

Weeks of school	9	18	27		
Weeks of vacation	4				
Total weeks	13				

b. Write a proportion.

$\frac{\square}{\square}$ $\frac{\square}{\square} = \frac{\square}{\square}$

c. Answer: ______

9. The class library has 3 fiction books for every 4 nonfiction books. If the library has a total of 63 books, how many fiction books does it have?

a. $\frac{\square}{\square}$ $\frac{\square}{\square} = \frac{\square}{\square}$

b. Answer: ______

Try This

10. There are 48 students in the sixth grade at Robert's school. Three out of 8 sixth graders read 2 books last month. One out of 3 students read only 1 book. The rest of the students read no books at all. How many books in all did the sixth graders read last month? ______

Tell what you did to solve the problem. ______

______

______

______

______

Date ____________ Time ____________

LESSON 8•7

Using Proportions to Solve Problems

Math Message

1. In a recent game, the Mansfield School basketball team took 15 three-point shots and made 6 of them. What percent of its shots did the team make? ________%

2. The team also took 20 two-point shots and made 45% of them. How many two-point shots did the players make? ________ two-point shots

3. The team made 80% of its free throws (one-point shots). If players made 16 free throws, how many free throws did they attempt? ________ free throws

4. How many shots did the team take in all? ________ shots

   How many points did the team score in all? ________ points

---

For each problem, use a variable to represent each part, whole, or percent that is unknown. Complete and solve each proportion.

5. Nigel's dog had a litter of puppies. Three of the puppies were male and 5 were female. What percent of the puppies were male?

   a. $\frac{\text{(male puppies)}}{\text{(total)}}$ $\frac{\square}{\square} = \frac{\square}{100}$

   b. ________% of the puppies were male.

6. The 12 boys in Mr. Stiller's class make up 40% of the class. How many students are in Mr. Stiller's class?

   a. $\frac{\text{(boys)}}{\text{(total)}}$ $\frac{\square}{\square} = \frac{\square}{100}$

   b. There are ________ students in Mr. Stiller's class.

7. Apples are about 85% water. What is the weight of the water in 5 pounds of apples?

   a. $\frac{\text{(weight of water)}}{\text{(weight of apples)}}$ $\frac{\square}{\square} = \frac{\square}{100}$

   b. There are ________ pounds of water in 5 pounds of apples.

Date Time

LESSON 8•7 Using Proportions to Solve Problems *cont.*

For each problem, use a variable to represent each part, whole, or percent that is unknown. Complete and solve each proportion.

8. 24 is what percent of 60?

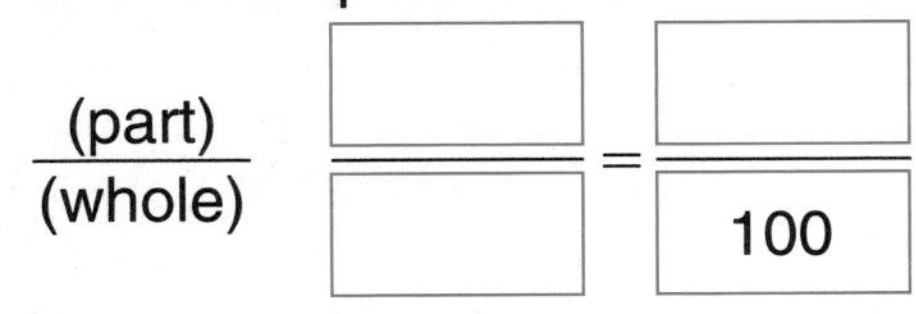

24 is ______% of 60.

9. 54 is what percent of 75?

(part) / (whole) ____ / ____ = ____ / ____

54 is ______% of 75.

10. 1 is what percent of 200?

(part) / (whole) ____ / ____ = ____ / ____

1 is ______% of 200.

11. 75% of what number is 24?

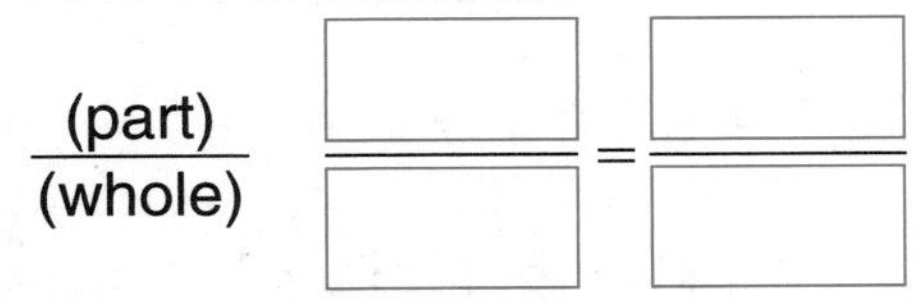

75% of ______ is 24.

12. 20 is 4% of what number?

(part) / (whole) ____ / ____ = ____ / ____

20 is 4% of ______.

Try This

Translate each question into an equation.

13. 75% of what number is 24?

Equation: ______________

14. 20 is 4% of what number?

Equation: ______________

15. Compare each equation you wrote to the cross products in Problems 11 and 12.

What do you notice? ______________

____

Date Time

LESSON 8•7 Math Boxes

1. A kilometer is about $\frac{5}{8}$ mile. About how many miles are in $4\frac{2}{5}$ kilometers? Set up a proportion below.

$\frac{\square}{\square} \quad \frac{\square}{\square} = \frac{\square}{\square}$

Solve the proportion in the space below.
Solution:

There are about ______ miles in $4\frac{2}{5}$ kilometers.

2. Solve.

Solution

a. $\frac{6}{24} = \frac{5}{n}$ ______

b. $\frac{p}{3.9} = \frac{100}{30}$ ______

c. $\frac{f}{\frac{7}{8}} = \frac{1\frac{1}{7}}{5}$ ______

3. Use order of operations to evaluate each expression.

a. $3 * 8 / 4 + 7 =$ ______

b. $9 + 3 * 5 - 7 =$ ______

c. ______ $= 6 * 5 + 7 * 3$

d. ______ $= 80 / (2 + 8) * 3^3 + 5$

e. ______ $= 28 - 7 * 4 * 0 + 2$

4. Which equation describes the relationship between the numbers in the table? Circle the best answer.

x	y
1	$\frac{3}{8}$
2	$\frac{3}{4}$
8	3
24	9

A. $\frac{y}{8} * 3 = x$

B. $(3 * y) + 8 = x$

C. $\frac{x}{8} * 3 = y$

D. $(3 * x) + 8 = y$

5. The area of triangle *FOG* is 12 cm^2. What is the perimeter of rectangle *FROG*?

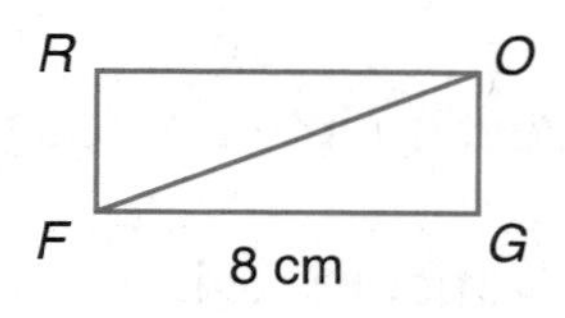

Perimeter = ______ cm

Date Time

LESSON 8·8

Math Boxes

1. Solve.

a. If \$933 is $\frac{1}{3}$ the original price, how much is the original price?

______

b. If $\frac{4}{9}$ of a box is 36 cookies, how many cookies are in the whole box?

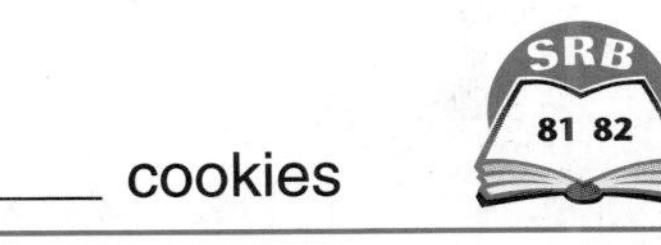

______ cookies

SRB 81 82

2. The ratio of managers to workers in a company is 3 to 11. If there are 42 managers, how many workers are there?

Write a proportion. Then solve.

There are ______ workers.

3. Peabody's Bookstore had a sale. Complete the table. Then use your protractor to make a circle graph of the information.

Book Category	Number Sold	Percent of Total	Degrees
Fiction	280		
Sports	283		
Children's	125		
Travel	212		
Computer	100		
Total			

______
(title)

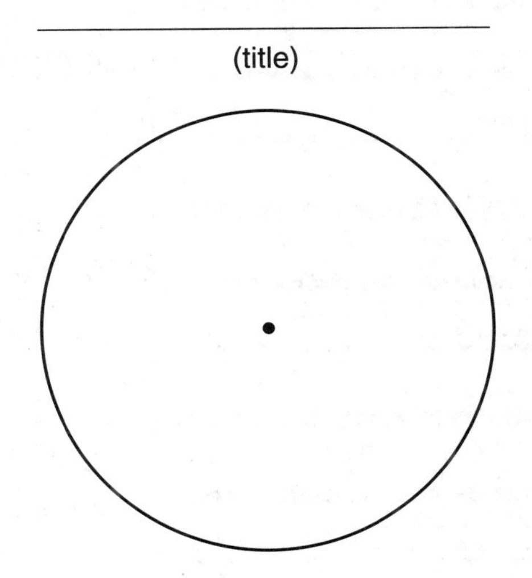

4. Use the general pattern

$x(y - z) = (x * y) - (x * z)$

to compute the following products mentally. One has been done for you.

$8 * 99 = 8(100 - 1) = (8 * 100) - (8 * 1) = 792$

a. $5 * 97 =$ ______

b. $12 * 18 =$ ______

5. The area A of a circle is given by the formula $A = \pi * r^2$, where r is the radius of the circle. Calculate the area of the circle below. Use $\pi = 3.14$.
Circle the best answer.

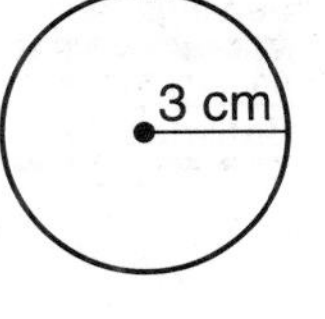

A. 12.14 cm^2

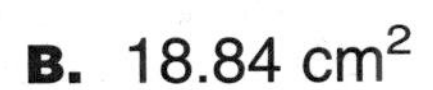

B. 18.84 cm^2

C. 19.72 cm^2

D. 28.26 cm^2

SRB 218

Date Time

LESSON 8•8 The Fat Content of Foods

1. Use the calorie information from the food labels on this page and the next.

a. Write the ratio of calories that come from fat to the total number of calories as a fraction.

b. Estimate the percent of total calories that come from fat. Do not use your calculator.

c. Use your calculator to find the percent of calories that come from fat. (Round to the nearest whole percent.)

Food Label	Food	$\frac{\text{Calories from Fat}}{\text{Total Calories}}$	Estimated Fat Percent	Calculated Fat Percent
Nutrition Facts Serving Size 1 slice (28 g) Servings Per Container 12 Amount Per Serving **Calories** 90 Calories from Fat 80	Bologna	$\frac{80}{90}$	About 90	89%
Nutrition Facts Serving Size 2 waffles (72 g) Servings Per Container 4 Amount Per Serving **Calories** 190 Calories from Fat 50	Waffle			
Nutrition Facts Serving Size 2 tablespoons (32 g) Servings Per Container 15 Amount Per Serving **Calories** 190 Calories from Fat 140	Peanut butter			
Nutrition Facts Serving Size 1 slice (19 g) Servings Per Container 24 Amount Per Serving **Calories** 70 Calories from Fat 50	American cheese			
Nutrition Facts Serving Size 1 egg (50 g) Servings Per Container 12 Amount Per Serving **Calories** 70 Calories from Fat 40	Egg			

Date Time

LESSON 8·8 The Fat Content of Foods *continued*

Food Label	Food	Calories from Fat / Total Calories	Estimated Fat Percent	Calculated Fat Percent
Nutrition Facts Serving Size 1 cup (60 mL) Servings Per Container 6 **Amount Per Serving** **Calories** 110 Calories from Fat 0	Orange juice			
Nutrition Facts Serving Size 1/2 cup (125 g) Servings Per Container About 3 1/2 **Amount Per Serving** **Calories** 90 Calories from Fat 5	Corn			
Nutrition Facts Serving Size 1 package (255 g) Servings Per Container 1 **Amount Per Serving** **Calories** 280 Calories from Fat 90	Macaroni and cheese			
Nutrition Facts Serving Size 1/2 cup (106 g) Servings Per Container 4 **Amount Per Serving** **Calories** 270 Calories from Fat 160	Ice cream			

2. Compare whole milk to skim (nonfat) milk.

Type of Milk	Total Calories	Calories from Fat	Calories from Carbohydrate	Calories from Protein
1 cup whole milk	160	75	50	35
1 cup skim milk	85	trace	50	35

a. For whole milk, what percent of the total calories come from

fat? ________% carbohydrate? ________% protein? ________%

b. For skim milk, what percent of the total calories come from

fat? ________% carbohydrate? ________% protein? ________%

3. Find the missing percents.

a. 25% + 30% + ________% = 100% **b.** 82% + ________% + 9% = 100%

Date Time

LESSON 8•9 Enlargements

A copy machine was used to make 2X enlargements of figures on the Geometry Template.

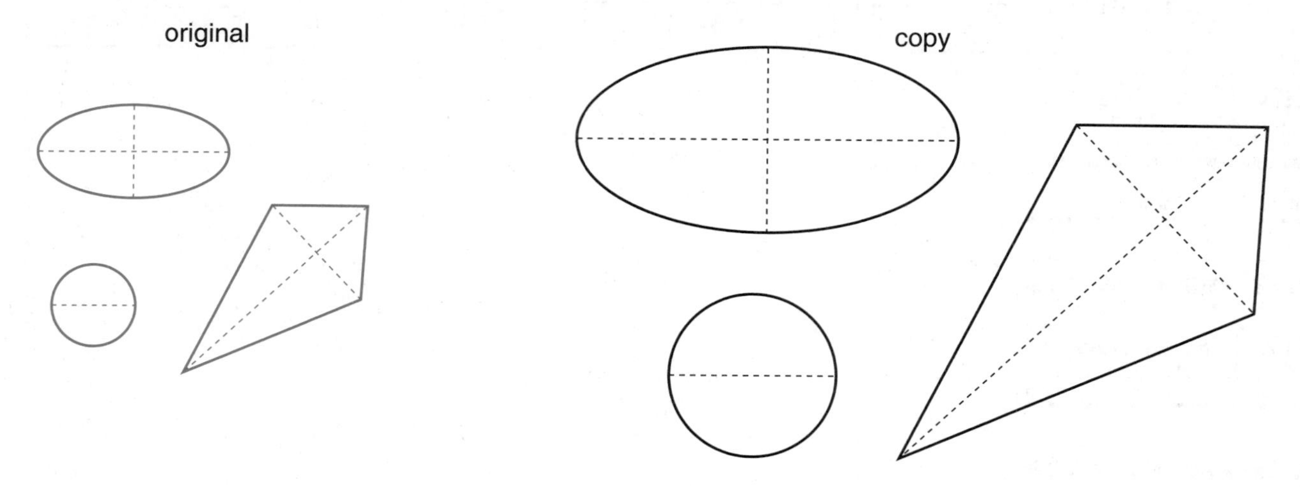

1. Use your ruler to measure the line segments shown in the figures above to the nearest $\frac{1}{16}$ inch. Then fill in the table below.

Line Segment	Length of Original	Length of Enlargement	Ratio of Enlargement to Original
Diameter of circle			
Longer axis of ellipse			
Shorter axis of ellipse			
Longer side of kite			
Shorter side of kite			
Longer diagonal of kite			
Shorter diagonal of kite			

2. Are the figures in the enlargements similar to the original figures? ______

3. What does a 3.5X enlargement mean? ______

______

______

______

Date Time

LESSON 8•9

Map Scale

This map shows the downtown area of the city of Chicago. The shaded area shows the part of Chicago that was destroyed in the Great Chicago Fire of 1871.

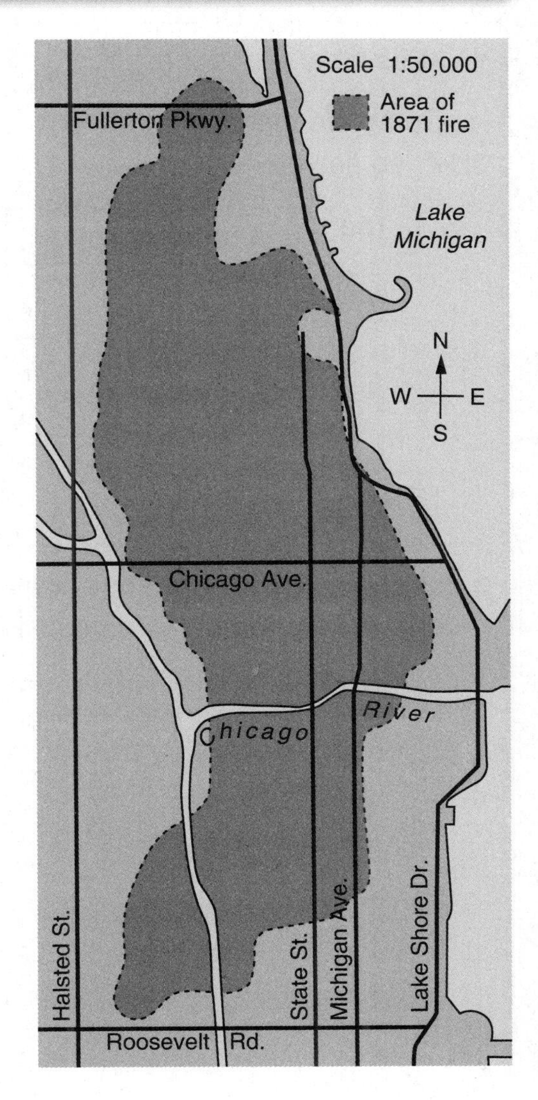

The map was drawn to a scale of 1:50,000. This means that each 1-inch length on the map represents 50,000 inches (about $\frac{3}{4}$ mile) of actual distance.

$$\frac{\text{map distance}}{\text{actual distance}} = \frac{1}{50,000}$$

1. Measure the distance on the map between Fullerton Parkway and Roosevelt Road, to the nearest $\frac{1}{4}$ inch. This is the approximate north–south length of the part that burned.

   Burn length on map = ______ inches

2. Measure the width of the part that burned along Chicago Avenue, to the nearest $\frac{1}{4}$ inch. This is the approximate east–west length of the part that burned.

   Burn width on map = ______ inches

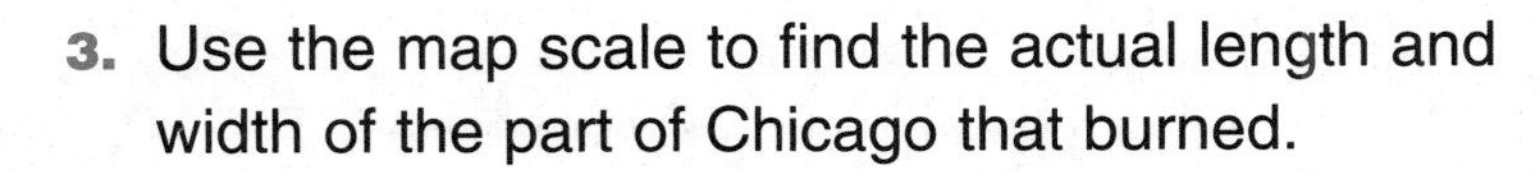

3. Use the map scale to find the actual length and width of the part of Chicago that burned.

   a. Actual burn length = ______________ inches

   b. Actual burn width = ______________ inches

4. Convert the answers in Problem 3 from inches to miles, to the nearest tenth of a mile.

   a. Actual burn length = ______ miles

   b. Actual burn width = ______ miles

5. Estimate the area of the part of Chicago that burned, to the nearest square mile.

   About ____________ square miles

Date Time

LESSON 8•9

Math Boxes

1. 15 is what percent of 25?

Complete the proportion. Then solve.

$$\frac{\text{(part)}}{\text{(whole)}} \quad \frac{\square}{\square} = \frac{\square}{\square}$$

15 is ________% of 25.

2. Find the value of x so each ratio is expressed in terms of a common unit.

a. 4 inches:5 feet = 4 inches:x inches

x = ________

b. $\frac{2.4 \text{ m}}{80 \text{ cm}} = \frac{2.4 \text{ m}}{x \text{ m}}$ x = ________

c. 840 mm to 7 cm = x cm to 7 cm

x = ________

3. Subtract. Write your answer as a fraction or a mixed number in simplest form.

a. $7\frac{3}{4} - 3\frac{3}{8}$ = ________

b. ________ $= \frac{5}{2} - 1\frac{5}{6}$

c. ________ $= 5\frac{1}{3} - 2\frac{5}{9}$

d. ________ $= 17 - 13\frac{4}{5}$

4. Write 5 names for the number 10 in the name-collection box. Each name should include the number (−2) and involve subtraction.

10

5. The spreadsheet shows how Jonas spent his money for the first quarter of the year.

a. In which cell is the largest amount that Jonas spent?

____

b. Calculate the values for cells E2, E3, and E4 and enter them in the spreadsheet.

	A	B	C	D	E
1	Month	Food	Movies	Music	Total
2	January	\$38.50	\$34.00	\$62.50	
3	February	\$29.45	\$28.70	\$26.89	
4	March	\$34.90	\$41.86	\$48.30	

c. Circle the correct formula for calculating the amount of money Jonas spent in February.

D1 + D2 + D3   D3 − C2 + C3   B3 + C3 + D3

Date Time

LESSON 8•10

Math Boxes

1. The owner of a restaurant knows that only about 75% of those who make a dinner reservation actually show up. At this rate, how many reservations should the owner take to fill his restaurant of 180 seats?

Write a proportion. Then solve.

Proportion: ______________________

Solution: ________

2. Triangles *JKL* and *PQR* are similar.

J, *K*, *L*: 8, 6, 10
P, *Q*, *R*: 12, *x*, 15

a. Find the ratio *JL*:*PR*. __________

b. The length of $\overline{QR}$ = ______

3. Rename the fraction to 2 decimal places.

$\frac{5}{7}$

$\frac{5}{7}$ rounded to the nearest hundredth = ________

4. Find the value of $x + y + z$.

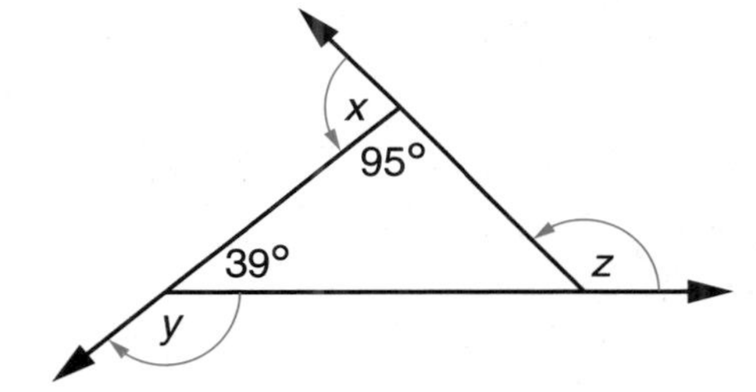

$x + y + z =$ ________

5. Which phrase is represented by the algebraic expression $4 - x$?
Fill in the circle next to the best answer.

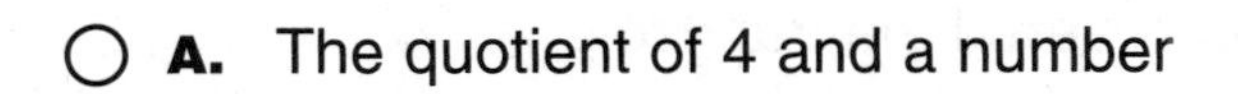

B. 4 less than a number

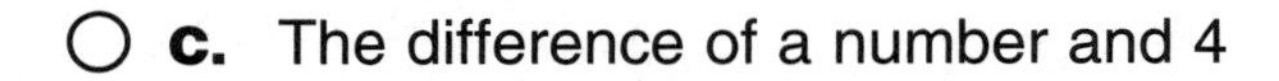

D. 4 decreased by a number

6. Write an equation to represent the congruent line segments shown below.

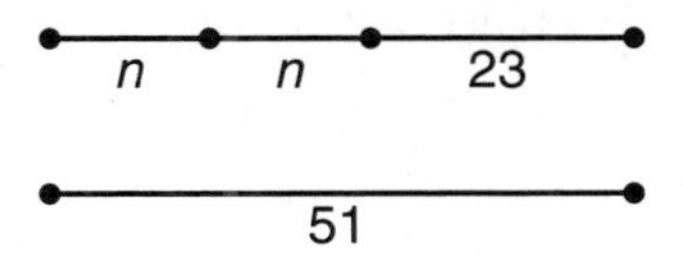

Equation: ______________________

Solve for *n*.

$n =$ ________

Similar Polygons

1. Use pattern-block trapezoids to construct a trapezoid whose sides are twice the length of the corresponding sides of a single pattern-block trapezoid. Then use your Geometry Template to record what you did.

2. Draw a trapezoid whose sides are 3 times the length of a single pattern-block trapezoid. You may use any drawing or measuring tools you wish, such as a compass, a ruler, a protractor, the trapezoid on your Geometry Template, or a trapezoid pattern block.

Which tools did you use? ______________________________

Try This

3. Cover the trapezoid you drew in Problem 2 with pattern-block trapezoids from the Geometry Template. Then record the way you covered the trapezoid.

Date Time

Similar Polygons *continued*

4. Measure line segments *AB, CD,* and *EF* to the nearest millimeter. Draw a line segment *GH* so the ratio of the lengths of *AB* to *CD* is equal to the ratio of the lengths of *EF* to *GH*.

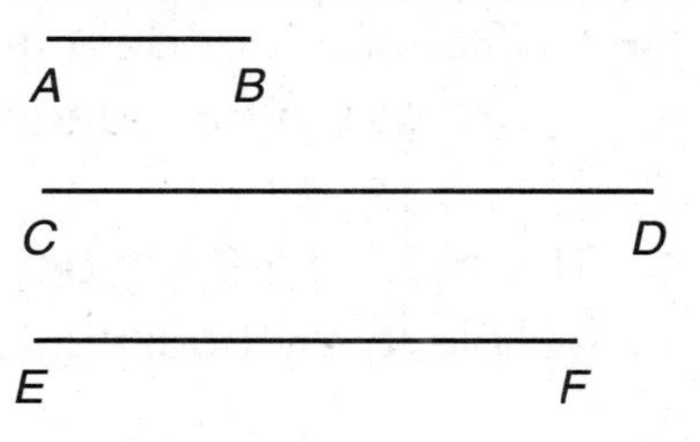

$$\frac{\text{length of } \overline{AB}}{\text{length of } \overline{CD}} = \frac{\text{length of } \overline{EF}}{\text{length of } \overline{GH}}$$

5. Pentagons *PAINT* and *MODEL* are similar polygons. Find the missing lengths of sides.

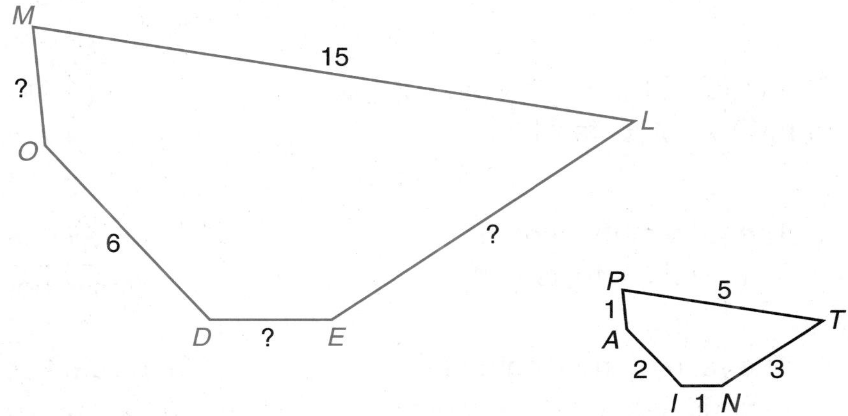

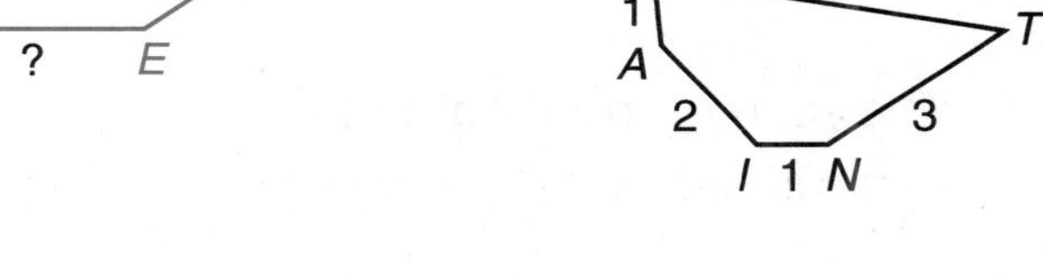

a. Length of side *MO* = _______ units

b. Length of side *EL* = _______ units

c. Length of side *DE* = _______ units

6. Triangles *PAL* and *CUT* are similar figures. Find the missing lengths of sides.

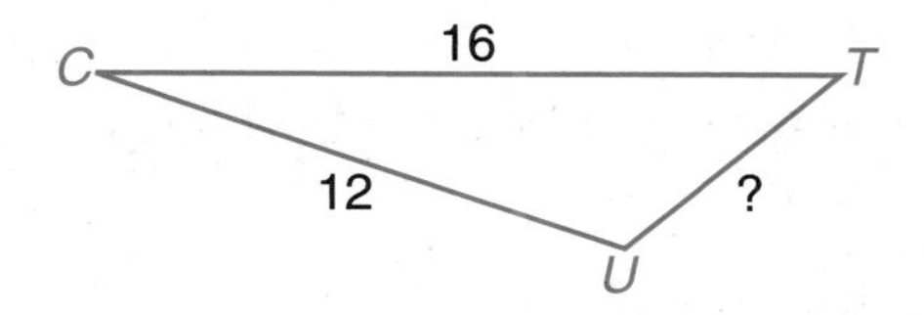

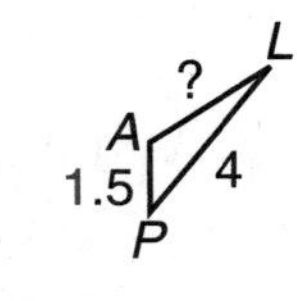

a. Length of side *AL* = _______ units

b. Length of side *UT* = _______ units

7. Alona is looking at a map of her town. The scale given on the map is 1 inch represents $\frac{1}{2}$ mile. Alona measures the distance from her home to school on the map. It's $3\frac{3}{4}$ inches. What is the actual distance from her home to school?

_______ miles

8. For a school fair in the cafeteria, Jariah wants to construct a scale model of the 984-foot Eiffel Tower. She plans to use a scale of 1 to 6. Every length of the scale model will be $\frac{1}{6}$ the actual size of the Eiffel Tower. Does this scale seem reasonable? If yes, explain why. If not, suggest a more reasonable scale.

____

____

Date Time

LESSON 8•11

Renaming and Comparing Ratios

Part 1 Record your data from Study Link 8-8 as *n*-to-1 ratios. Use your calculator to divide. Round to the nearest tenth.

1. The ratio of left-handed to right-handed people in my household $\frac{\text{(left-handed)}}{\text{(right-handed)}}$ $\frac{\square}{\square}$ is about $\frac{\square}{1}$

2. The ratio of the length of the American flag I found to its width $\frac{\text{(flag length)}}{\text{(flag width)}}$ $\frac{\square}{\square}$ is about $\frac{\square}{1}$

3. The ratio of the length of the screen of my TV set to its width $\frac{\text{(TV length)}}{\text{(TV width)}}$ $\frac{\square}{\square}$ is about $\frac{\square}{1}$

4. **a.** The ratio of the length of a small book to its width $\frac{\text{(small book length)}}{\text{(small book width)}}$ $\frac{\square}{\square}$ is about $\frac{\square}{1}$

   b. The ratio of the length of a medium book to its width $\frac{\text{(medium book length)}}{\text{(medium book width)}}$ $\frac{\square}{\square}$ is about $\frac{\square}{1}$

   c. The ratio of the length of a large book to its width $\frac{\text{(large book length)}}{\text{(large book width)}}$ $\frac{\square}{\square}$ is about $\frac{\square}{1}$

   d. What is the shape of a book with a length-to-width ratio of 1 to 1? ____________________

5. **a.** The ratio of the length of a postcard to its width $\frac{\text{(postcard length)}}{\text{(postcard width)}}$ $\frac{\square}{\square}$ is about $\frac{\square}{1}$

   b. The ratio of the length of an index card to its width $\frac{\text{(index card length)}}{\text{(index card width)}}$ $\frac{\square}{\square}$ is about $\frac{\square}{1}$

   c. The ratio of the length of a regular-size envelope to its width $\frac{\text{(envelope length)}}{\text{(envelope width)}}$ $\frac{\square}{\square}$ is about $\frac{\square}{1}$

   d. The ratio of the length of a business envelope to its width $\frac{\text{(business envelope length)}}{\text{(business envelope width)}}$ $\frac{\square}{\square}$ is about $\frac{\square}{1}$

   e. The ratio of the length of a sheet of notebook paper to its width $\frac{\text{(notebook paper length)}}{\text{(notebook paper width)}}$ $\frac{\square}{\square}$ is about $\frac{\square}{1}$

Date _______________ Time _______________

LESSON 8·11

Renaming and Comparing Ratios *cont.*

6. Measure the length and width of each rectangle in Problem 6 on Study Link 8-8 to the nearest millimeter. Find the ratio of length to width for each rectangle.

a. $\frac{\text{(length of A)}}{\text{(width of A)}}$ $\frac{\square}{\square} = \frac{\square}{1}$

b. $\frac{\text{(length of B)}}{\text{(width of B)}}$ $\frac{\square}{\square} = \frac{\square}{1}$

c. $\frac{\text{(length of C)}}{\text{(width of C)}}$ $\frac{\square}{\square} = \frac{\square}{1}$

d. $\frac{\text{(length of D)}}{\text{(width of D)}}$ $\frac{\square}{\square} = \frac{\square}{1}$

e. Which of the four rectangles was the most popular? _______________

7. The ratio of the rise to the run of my stairs is $\frac{\text{(rise)}}{\text{(run)}}$ $\frac{\square}{\square} = \frac{\square}{1}$

Part 2 Share the data you recorded in Problems 1–7 with the other members of your group. Use these data to answer the following questions.

8. Which group member has the largest ratio of left-handed people to right-handed people at home? _______________ What is this ratio? _______________

9. By law, the length of an official American flag must be 1.9 times its width.

a. Did the flag you measured meet this standard? _______________

b. What percent of the flags measured by your group meet this standard? _______________

c. One of the largest American flags was displayed at the J. L. Hudson store in Detroit, Michigan. The flag was 235 feet by 104 feet. Does this flag meet the legal requirements? _______________

How can you tell? _______________

____

10. For standard television sets, the ratio of length to width of the screen is about 4 to 3. For widescreen TVs, this ratio is about 16:9.

a. Is this true for the television sets in your group? _______________

b. Which television screen, standard or widescreen, is closest to having an *n*-to-1 ratio of 1.6?

____

Date Time

LESSON 8•11

Renaming and Comparing Ratios *cont.*

11. Compare the ratios for the small, medium, and large books from Problem 4 that were measured by your group.

Which size of books tends to have the largest ratio of length to width? ______________

Which size tends to have the smallest ratio? ______________

12. It is often claimed that the nicest looking rectangular shapes have a special ratio of length to width. Such rectangles are called **Golden Rectangles.** In a Golden Rectangle, the ratio of length to width is about 8 to 5.

A B C D

a. Which of the four rectangles—A, B, C, or D—is closest to having the shape of a Golden Rectangle? ________

b. Did most people in your family choose the Golden Rectangle? ____________

c. Draw a Golden Rectangle in the space at the right whose shorter sides are 2 centimeters long.

d. Which TV screen in Problem 10 on page 317 is closest to a Golden Rectangle? ______________________

13. Most stairs in homes have a rise of about 7 inches and a run of about $10\frac{1}{2}$ inches. Therefore, the rise is about $\frac{2}{3}$ the run.

a. Is this true of your stairs? ____________

b. Which stairs would be steeper, stairs with a rise-to-run ratio of 2:3 or 3:2?

____

c. Which member of your group has the steepest stairs? ____________

What is the ratio of rise to run? ____________

d. On the grid at the right, draw stairs whose rise is $\frac{2}{5}$ the run.

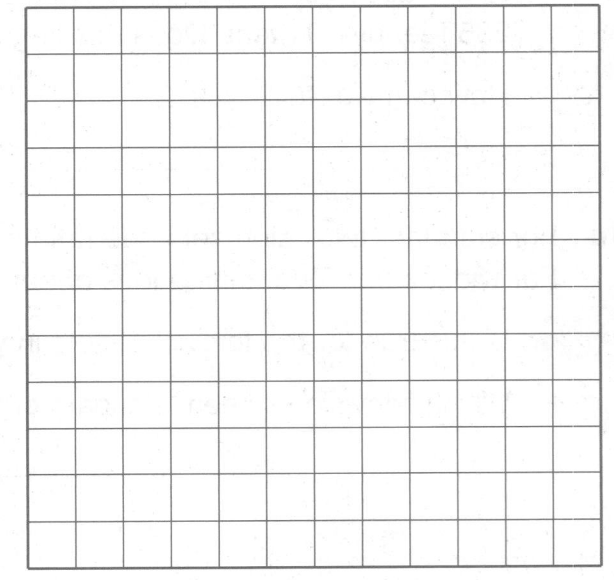

Date Time

LESSON 8•11

Math Boxes

1. What percent of 56 is 14?

Complete the proportion.
Then solve.

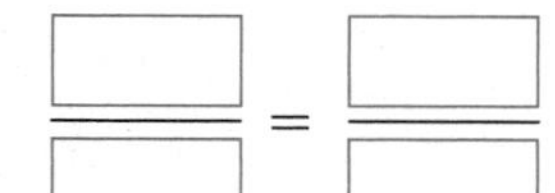

14 is ________% of 56.

2. Find the value of x so each ratio is expressed in terms of a common unit.

a. 9 inches:4 yards = x yards:4 yards

x = ________

b. $\frac{6 \text{ hours}}{3 \text{ days}} = \frac{x \text{ days}}{3 \text{ days}}$ x = ________

c. 140 quarts to 560 pints =
140 quarts to x quarts

x = ________

3. Subtract. Write your answer as a fraction or a mixed number in simplest form.

a. $\frac{3}{2} - \frac{5}{8}$ = ________

b. ________ $= 4\frac{2}{3} - 1\frac{1}{2}$

c. ________ $= 3\frac{1}{4} - 1\frac{5}{6}$

d. $5\frac{8}{9} - \frac{25}{25}$ = ________

4. Write 5 names for the number 8 in the name-collection box. Each name should include the fraction $\frac{1}{3}$ and involve multiplication.

8

5. The spreadsheet shows Hiroshi's utility bills for 2 months.

a. If Hiroshi entered the wrong amount for the January electric bill, which cell should he correct?

____

b. Calculate the values for cells E2 and E3 and enter them in the spreadsheet.

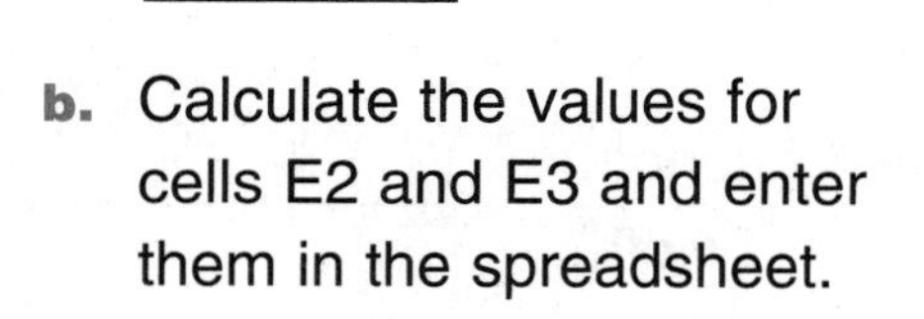

	A	B	C	D	E
1	Month	Phone	Electric	Gas	Total
2	January	\$17.95	\$38.50	\$120.50	
3	February	\$34.70	\$35.60	\$148.96	

c. Circle the correct formula for calculating the total cost of the utilities for February.

A2 + B2 + C2 + D2 B3 + C3 + D3 (B3 + C2 + D2) / 3

Date Time

LESSON 8•12

Rectangle Length-to-Width Ratios

Math Message

1. Draw a large rectangle.

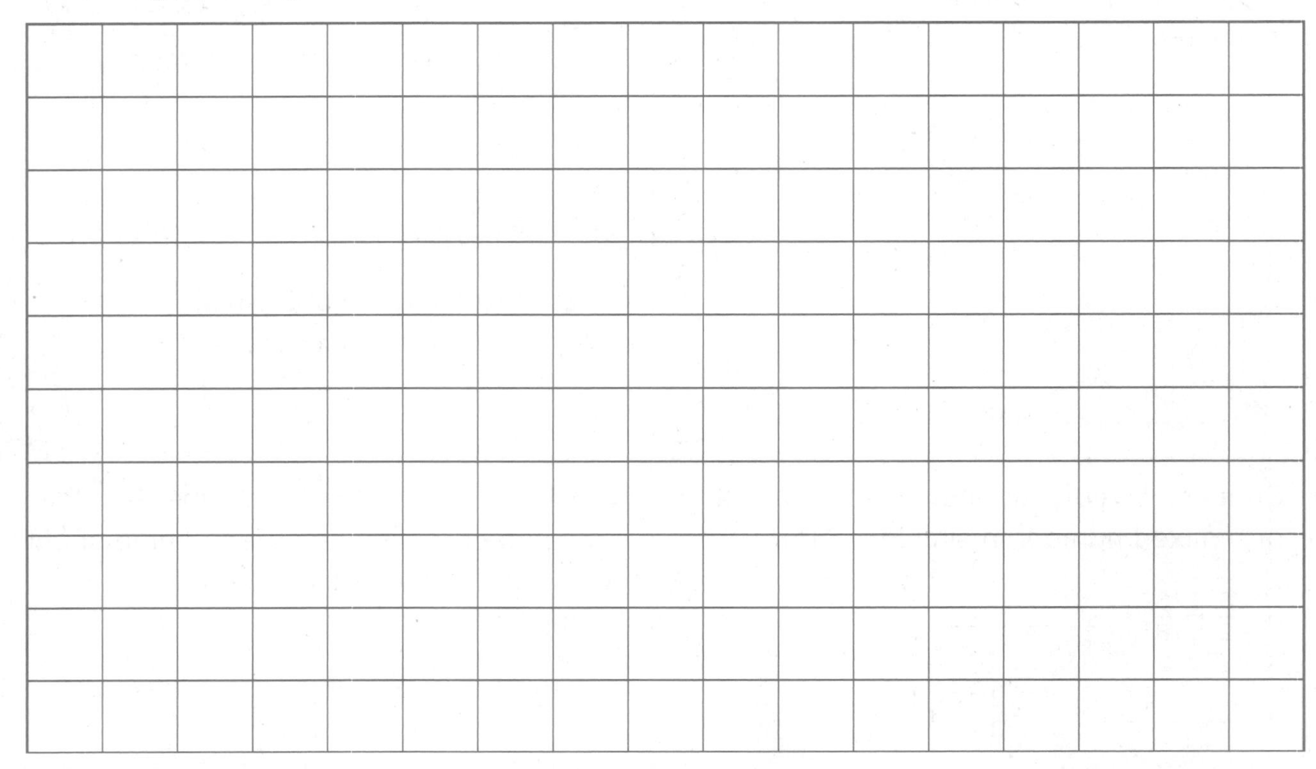

2. Measure the length and width of your rectangle to the nearest millimeter. Calculate the ratio of length to width. (Call the longer side the length and the shorter side the width.)

$\frac{\text{(length)}}{\text{(width)}} \quad \frac{\square}{\square} = \frac{\square}{1}$

3. Using a compass, draw 2 arcs on your rectangle as shown at the right. The arcs are drawn with the compass point at vertices that are next to each other. The compass opening is the same for both arcs.

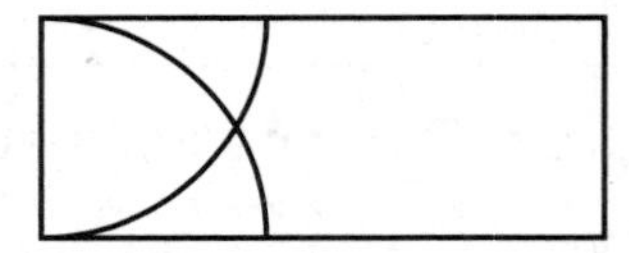

4. Connect the ends of your arcs to make a square. Shade the square. Your rectangle should now look something like this:

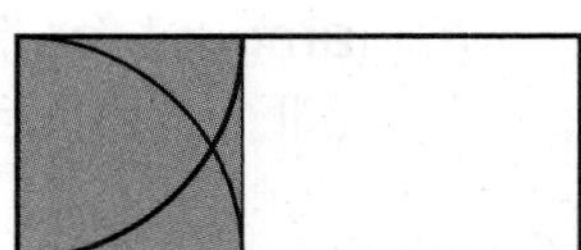

5. Measure the length and width of the unshaded part of your rectangle to the nearest millimeter. Calculate the ratio of the length of the rectangle to its width. (As before, call the longer side the length and the shorter side the width.)

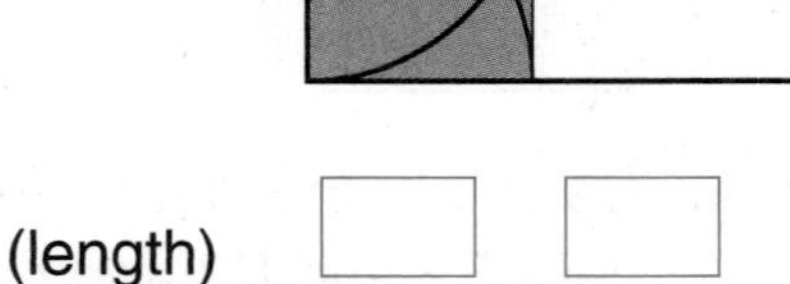

6. Are the ratios you calculated in Problems 2 and 5 equal? ________

LESSON 8•12

Ratios in a Golden Rectangle

1. Measure the length and width of rectangle *ABCD* to the nearest millimeter. (Call the longer side the length and the shorter side the width.) Calculate the length-to-width ratio to the nearest tenth.

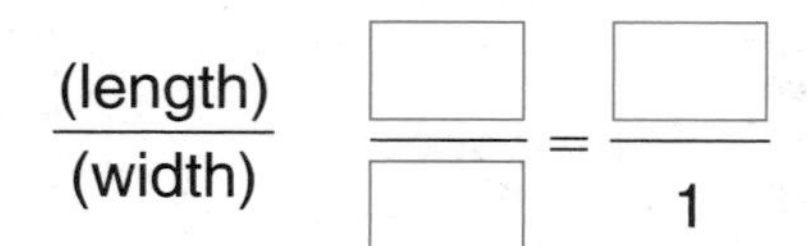

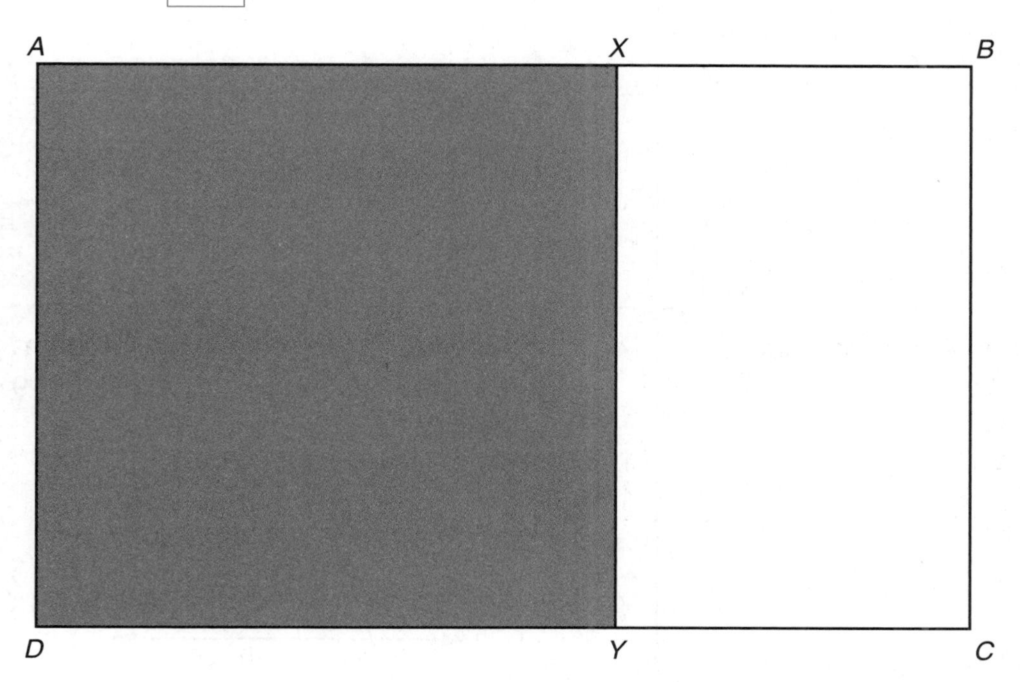

2. Measure the sides of rectangle *AXYD.* What kind of rectangle is *AXYD?*

____

3. Measure the length and width of rectangle *XBCY.* (Call the longer side the length and the shorter side the width.) Calculate the length-to-width ratio.

$\frac{\text{(length)}}{\text{(width)}}$ $\frac{\square}{\square} = \frac{\square}{1}$

4. What do you notice about the ratios you calculated in Problems 1 and 3?

____

____

Date Time

LESSON 8•12

Math Boxes

1. A scale drawing of a giraffe is 2% of actual size. If the drawing is 12 cm high, what is the actual height of the giraffe?

Write a proportion. Then solve.

Proportion: ______________________

Solution: ______________

2. The snapshot is a reduction of the poster. Find the poster's width. Show your work.

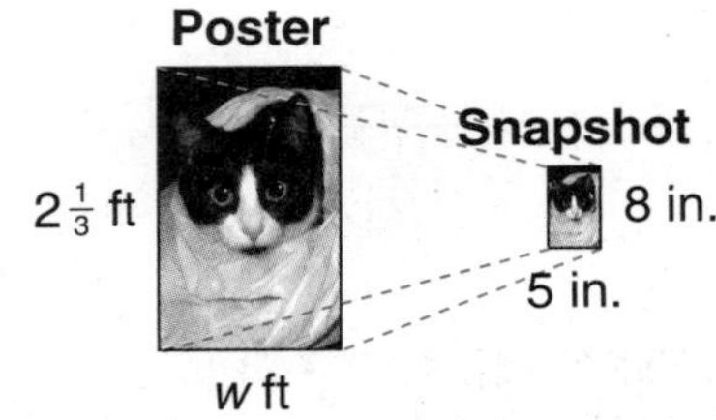

$w =$ __________

3. Rename the fraction to 2 decimal places.

$\frac{9}{13}$

$\frac{9}{13}$ rounded to the nearest hundredth = __________

4. Without using a protractor, find the sum of the angles numbered 1, 3, 5, 7, and 9. (Lines *a* and *b* are parallel.)

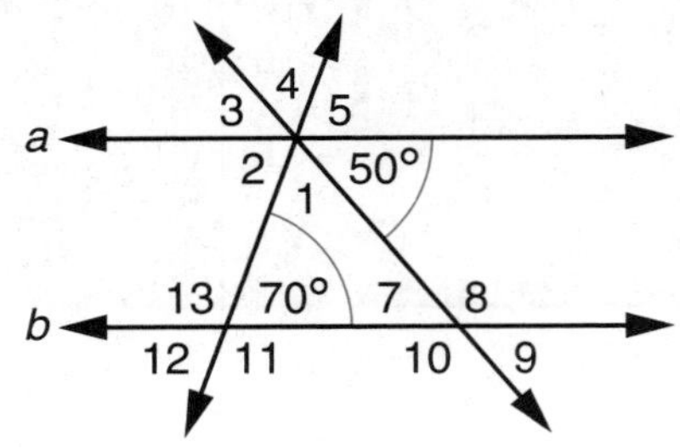

$m\angle 1 + m\angle 3 + m\angle 5 + m\angle 7 + m\angle 9 =$

____

5. Write an algebraic expression for the area of the rectangle below. Use $A = b * h$.

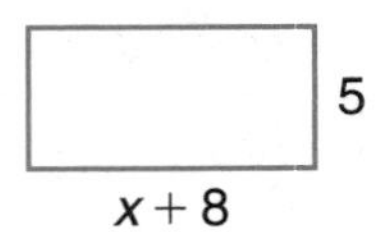

Expression: ______________________

Find the area of the rectangle when $x = 3$.

Area = __________ units^2

6. Write an equation to find the value of *m*.

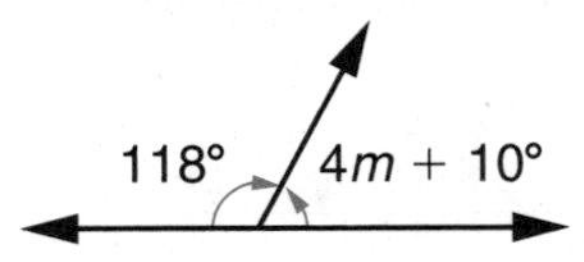

Equation: ______________________

Solve for *m*.

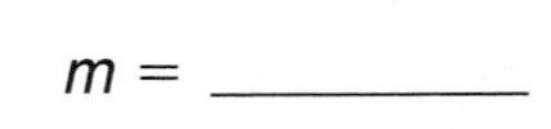

$m =$ __________

Date Time

LESSON 8•13 Math Boxes

1. Insert parentheses to make each equation true.

a. $\frac{1}{2} * 18 + 2 * 15 = 150$

b. $9 + 7 \div 4 * 2 = 8$

c. $5 / 3 + 3 * 5 = 4\frac{1}{6}$

d. $6 \div 17 - 11 * 14 = 14$

e. $4 / 9 + 3 * 6 = 2$

2. The circumference of a circle is given by the formula $C = \pi * d$, where C is the circumference and d is the diameter. Circle an equivalent formula.

$d = \frac{C}{\pi}$ $d = C * \pi$

$d = \frac{\pi}{C}$ $d = \pi + C$

Find the circumference for a circle with a diameter of 5 cm. Use 3.14 for π.

$C =$ ________ cm

3. Mr. and Mrs. Gauss keep a record of their expenses on a spreadsheet.

	A	B	C
1	Total of Expense	June	July
2	Rent and Utilities	$755	$723
3	Food	$125	$189
4	Car Expenses	$179	$25
5	Clothing	$65	$0
6	Miscellaneous	$45	$23
7	Total		

a. If the Gausses entered the wrong amount for car expenses in July, which cell should they correct? ________

b. In which month were the total expenses greater? ____________

How much greater? ____________

c. Circle the correct formula for the total for June.

A7 + B7 + C7 B2 + B3 + B4 + B5 + B6 (B2 + B3 + B4 + B5 + B6) / 5

4. Solve the equation.

$-47 + 6k = 3k - 2$

$k =$ ________

5. Add or subtract.

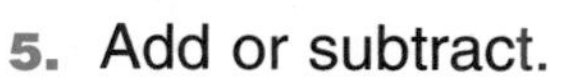

a. $235 + (-150) =$ ____________

b. $-76 - 24 =$ ____________

c. ____________ $= 143 - 258$

d. ____________ $= -99 + 167$

e. ____________ $= 380 - (-59)$

Date ____________ Time ____________

LESSON 9•1 Two Methods for Finding Areas of Rectangles

Math Message

Rectangle A

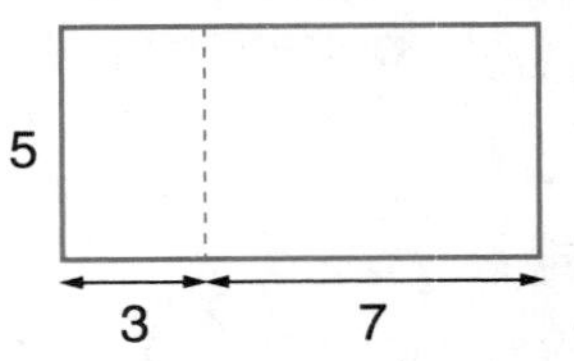

1. What is the area of Rectangle A? ________ square units

   We can express the area of Rectangle A in 4 ways.

   $5 * (3 + 7)$ $(5 * 3) + (5 * 7)$

   $(3 + 7) * 5$ $(3 * 5) + (7 * 5)$

Rectangle B

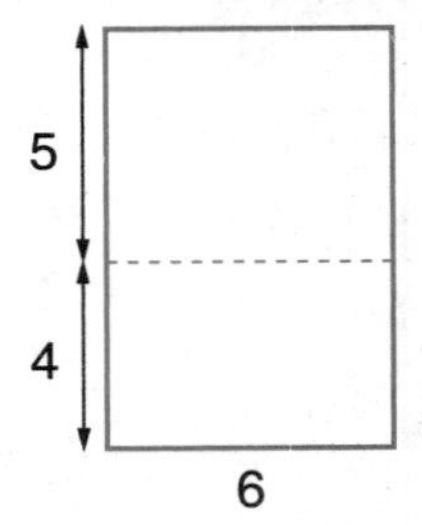

2. Write 2 number sentences to express the area of Rectangle B.

   ________ * (________ + ________) = ________

   (________ * ________) + (________ * ________) = ________

Rectangle C

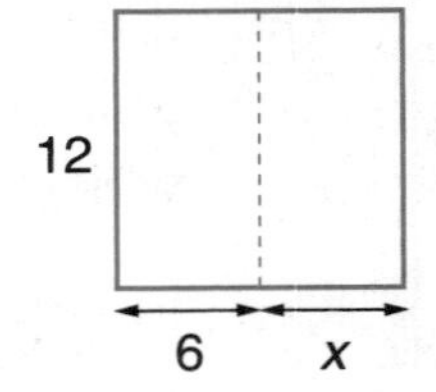

3. The area of Rectangle C is 144 square units.

   a. What is the value of *x*? ________

   b. Write 2 number sentences to express the area of Rectangle C.

   ________ * (________ + x) = 144

   (________ * ________) + (________ * x) = 144

4. Each of the following expressions describes the area of one of the rectangles below. Write the letter of the rectangle next to the correct expression. The first one has been done for you.

Rectangle D

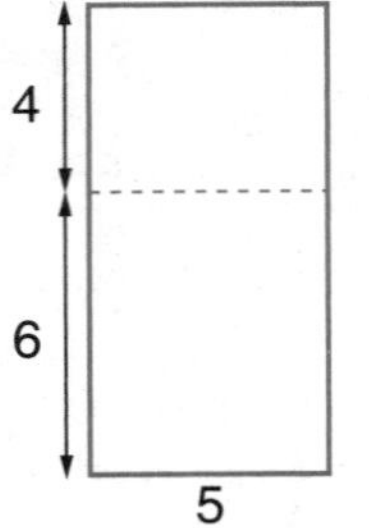

Rectangle E

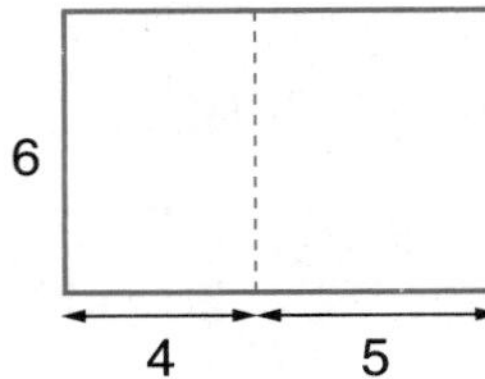

Rectangle F

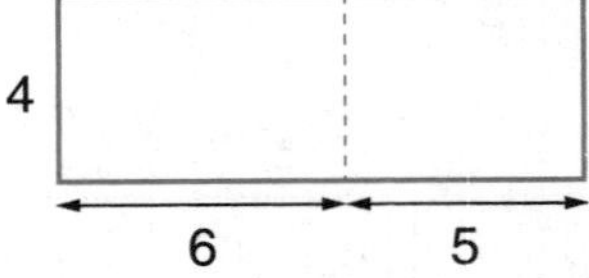

a. $6 * (5 + 4)$ ___E___

b. $(4 + 6) * 5$ ________

c. 44 ________

d. $24 + 30$ ________

e. $(6 * 4) + (5 * 4)$ ________

f. 50 ________

g. $(5 * 6) + (4 * 6)$ ________

h. $24 + 20$ ________

i. $(6 + 5) * 4$ ________

j. $(5 * 6) + (5 * 4)$ ________

Date ______________________ Time ______________________

LESSON 9·1 Two Methods for Finding Areas of Rectangles *cont.*

5. What is the area of the shaded part of Rectangle G?

Area of shaded part = ________ square units

We can express the area of the shaded part of Rectangle G with a number sentence in 4 ways.

$5 * (10 - 7) = 15$ $(5 * 10) - (5 * 7) = 15$

$(10 - 7) * 5 = 15$ $(10 * 5) - (7 * 5) = 15$

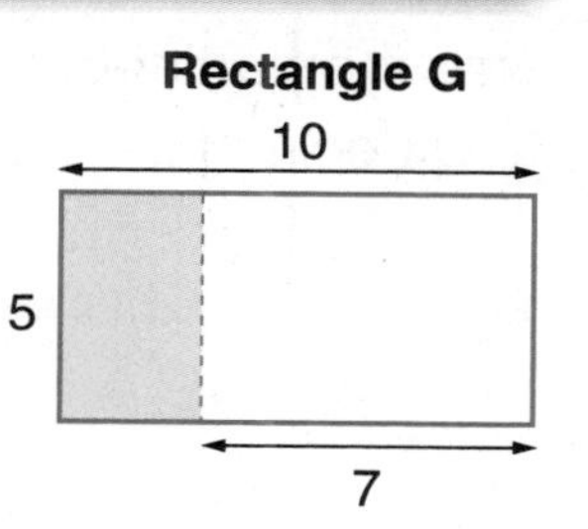

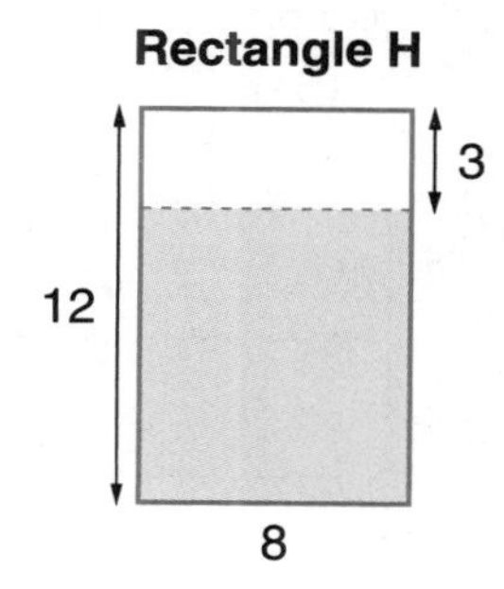

6. Write 2 number sentences to express the area of the shaded part of Rectangle H.

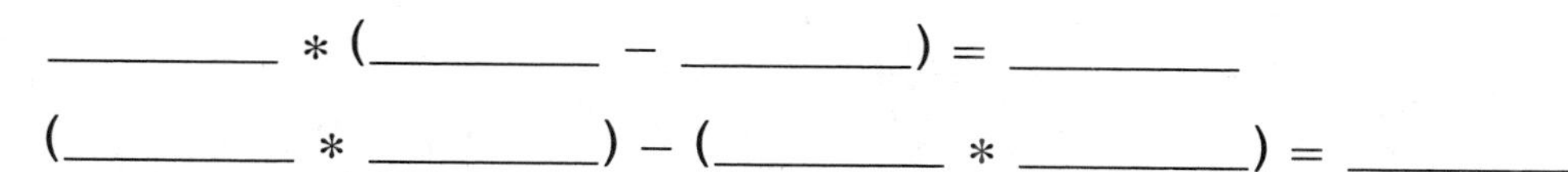

________ * (________ – ________) = ________

(________ * ________) – (________ * ________) = ________

7. The area of Rectangle I is 48 square units.

a. What is the value of y? ________

b. Write 2 number sentences to express the area of the shaded part of Rectangle I.

(________ – ________) $* y = 30$

(________ $* y) -$ (________ $* y) = 30$

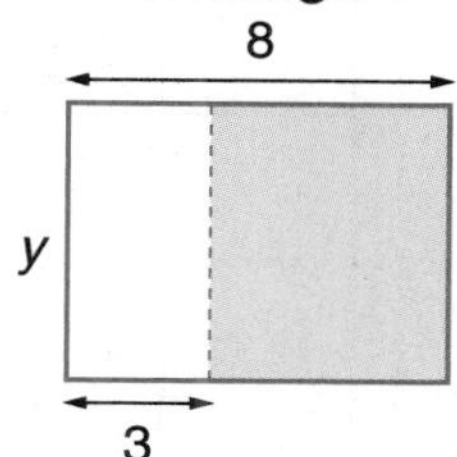

8. Each of the following expressions describes the area of the shaded part of one of the rectangles below. Write the letter of the rectangle next to the correct expression.

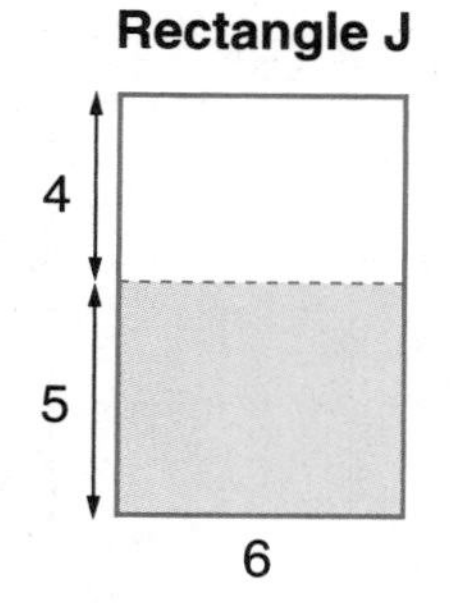

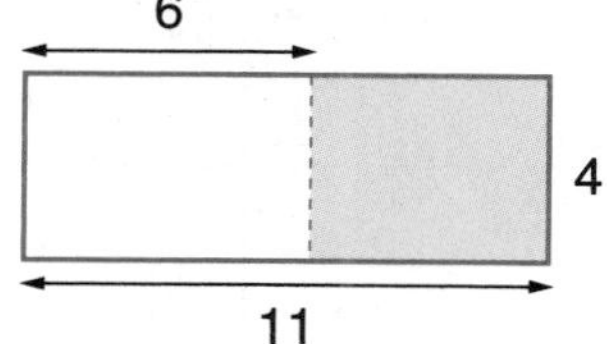

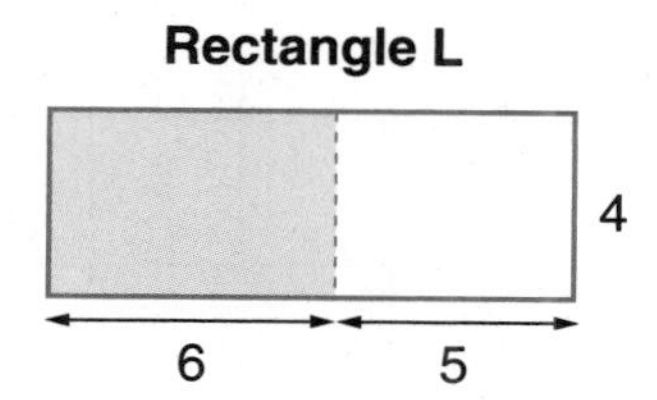

a. $4 * (11 - 6)$ ________

b. $44 - 20$ ________

c. 30 ________

d. $(6 * 9) - (6 * 4)$ ________

e. $(4 * 11) - (4 * 6)$ ________

f. $(11 - 5) * 4$ ________

g. $(11 * 4) - (5 * 4)$ ________

h. $6 * (9 - 4)$ ________

Date _______________ Time _______________

LESSON 9·1

Partial-Quotients Division

Use the partial-quotients division algorithm to find quotients that are correct to 2 decimal places. Show your work on the computation grid below.

1. $\frac{1,285}{7}$ _______________

2. 3,709 ÷ 18 _______________

3. 42)$\overline{7,956}$ _______________

4. $\frac{282.25}{16}$ _______________

5. 19.015 ÷ 38 _______________

6. 3.8)$\overline{746.85}$ _______________

Date Time

LESSON 9•1 Math Boxes

1. Write 2 different number sentences for the area of the shaded part of the rectangle.

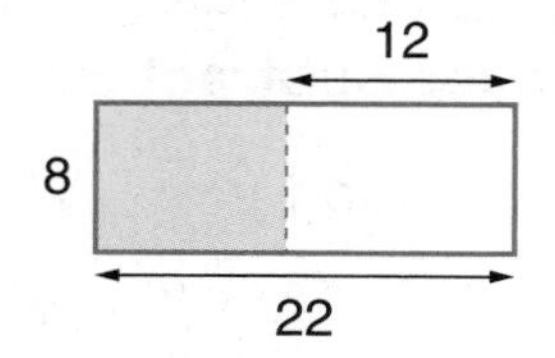

a. (______ − ______) ∗ ______ = 80

b. (______ ∗ ______) − (______ ∗ ______) = 80

2. Solve.

a. $15 = t - 6$

Solution ______

b. $y - (-12) = -7$

Solution ______

c. $6 + 10k = 256$

Solution ______

3. Triangles *QRS* and *XYZ* are similar.

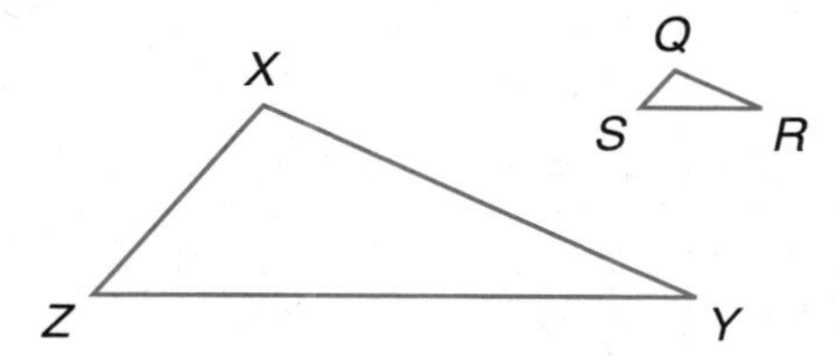

a. Find *SQ* when $SR = 7$, $ZX = 15$, and $ZY = 35$. $SQ =$ ______ units

b. Find the size-change factor: $\frac{\text{triangle } XYZ}{\text{triangle } QRS}$ = ______:______

4. Circle all the regular polygons.

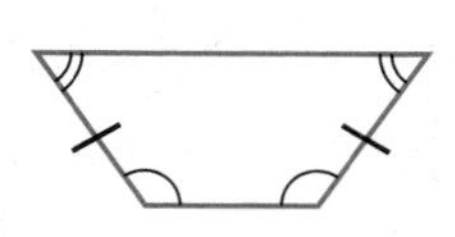

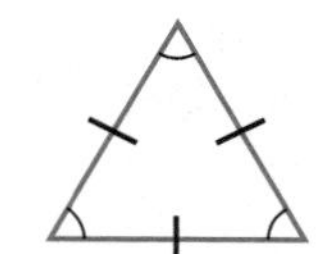

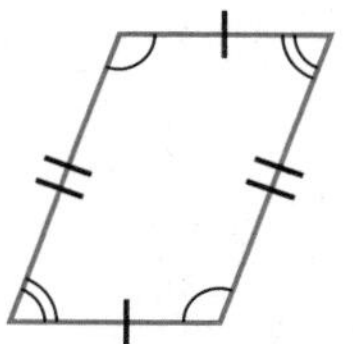

5. Use the diagonals to find the sum of the interior angle measures of the pentagon shown below.

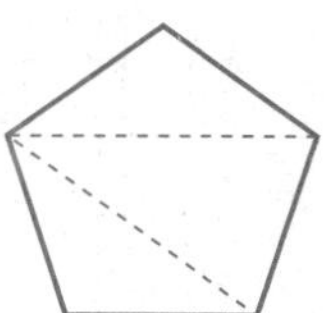

Sum of the interior angle measures = ______°

SRB 233

Date Time

LESSON 9·2

The Distributive Property

The distributive property is a number property that combines multiplication with addition or multiplication with subtraction. The distributive property can be stated in 4 different ways.

Multiplication over Addition	Multiplication over Subtraction
For any numbers a, x, and y:	For any numbers a, x, and y:
$a * (x + y) = (a * x) + (a * y)$	$a * (x - y) = (a * x) - (a * y)$
$(x + y) * a = (x * a) + (y * a)$	$(x - y) * a = (x * a) - (y * a)$

Use the distributive property to fill in the blanks.

1. $4 * (70 + 8) = (4 * \_\_\_\_\_) + (4 * \_\_\_\_\_)$
2. $6 * 34 = (\_\_\_\_\_ * 30) + (\_\_\_\_\_ * 4)$
3. $(6 * 70) - (6 * 4) = \_\_\_\_\_ * (70 - \_\_\_\_\_)$
4. $(\_\_\_\_\_ + \_\_\_\_\_) * 8 = (40 * 8) + (6 * \_\_\_\_\_)$
5. $8 * (90 + 3) = (\_\_\_\_\_ * 90) + (8 * 3)$
6. $(50 * 7) + (8 * \_\_\_\_\_) = (\_\_\_\_\_ + \_\_\_\_\_) * 7$
7. $9 * (20 - 7) = (9 * \_\_\_\_\_) - (\_\_\_\_\_ * 7)$
8. $4 * (5 + 6) = (\_\_\_\_\_ * \_\_\_\_\_) + (\_\_\_\_\_ * \_\_\_\_\_)$
9. $(41 + 19) * 7 = (\_\_\_\_\_ * \_\_\_\_\_) + (\_\_\_\_\_ * \_\_\_\_\_)$
10. $(18 - 4) * r = (18 * \_\_\_\_\_) - (\_\_\_\_\_ * r)$
11. $7 * (w - \_\_\_\_\_) = (\_\_\_\_\_ * w) - (\_\_\_\_\_ * 6)$
12. $n * (13 - 27) = (\_\_\_\_\_ * \_\_\_\_\_) - (\_\_\_\_\_ * \_\_\_\_\_)$
13. $(f - 8) * 15 = (\_\_\_\_\_ * \_\_\_\_\_) - (\_\_\_\_\_ * \_\_\_\_\_)$
14. $(29 * x) + (12 * x) = (\_\_\_\_\_ + \_\_\_\_\_) * \_\_\_\_\_$
15. $6 * (d - 7) = \_\_\_\_\_\_\_\_\_\_\_\_\_\_\_\_\_\_\_\_$
16. $5 * (12 - h) = \_\_\_\_\_\_\_\_\_\_\_\_\_\_\_\_\_\_\_\_$

Date Time

LESSON 9•2

Math Boxes

1. Use the distributive property to fill in the blanks.

a. $8 * (30 + 4) =$

$(\_\_\_\_\_ * \_\_\_\_\_) + (\_\_\_\_\_ * \_\_\_\_\_)$

b. $(\_\_\_\_\_ * 7) + (\_\_\_\_\_ * 6) = 9 * (7 + 6)$

c. $(20 + 6) * 10 = (20 * 10) + (6 * \_\_\_\_\_)$

d. $\_\_\_\_\_ (9 + 12) = (5)(9) + (5)(12)$

2. Circle the expressions that represent the area of the rectangle.

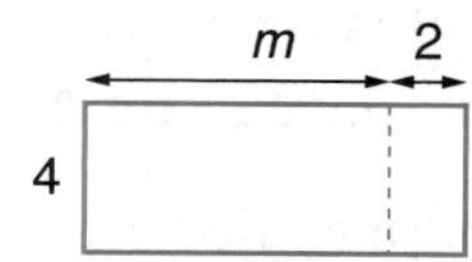

a. $4m + 8$

b. $4 * 2m$

c. $8m$

d. $(m + 2) * 4$

e. $4(2 + m)$

f. $8 + 2m$

3. Find the number.

a. $\frac{1}{10}$ of what number is 17? _____

b. $\frac{3}{4}$ of what number is 75? _____

c. $\frac{2}{5}$ of what number is 14? _____

d. $\frac{3}{20}$ of what number is 9? _____

e. $\frac{7}{8}$ of what number is 84? _____

4. Write >, <, or =.

a. $28 + (-15)$ _____ $36 \div (-2)$

b. $\frac{1}{2} + (-\frac{3}{4})$ _____ $\frac{2}{3} * \frac{7}{8}$

c. $-400 * -3$ _____ 20^2

d. $2 + 15 / 3$ _____ $7 * 10^{-1}$

e. $\frac{3}{7} + 6\frac{2}{3}$ _____ $\frac{12}{2} \div \frac{7}{9}$

5. Plot and label points on the coordinate grid as directed.

a. Plot (4,−2). Label it *A*.

b. Plot (−4,2). Label it *B*.

c. Draw line segment *AB*.

d. Name the coordinates of the midpoint of $\overline{AB}$.

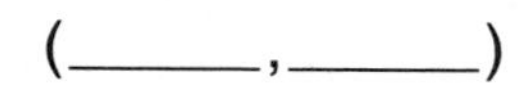

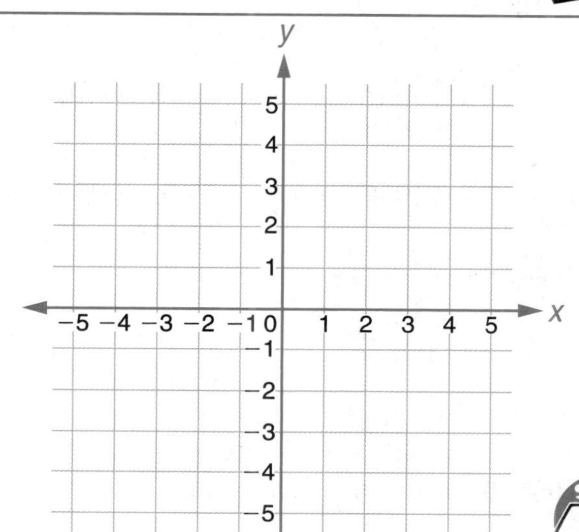

SRB 234

Date _______________ Time _______________

LESSON 9·3 Combining Like Terms

Algebraic expressions contain **terms.** For example, the expression $4y + 2x - 7y$ contains the terms $4y$, $2x$, and $7y$. The terms $4y$ and $7y$ are called **like terms** because they are multiples of the same variable, y. To **combine like terms** means to rewrite the sum or difference of like terms as a single term. In the case of $4y + 2x - 7y$, the like terms $4y$ and $-7y$ can be combined and rewritten as $-3y$.

To **simplify an expression** means to write the expression in a simpler form. Combining like terms is one way to do that. *Reminder:* The multiplication symbol ($*$) is often not written. For example, $4 * y$ is often written as $4y$, and $(x + 3) * 5$ as $(x + 3)5$.

Example 1: Simplify the expression $5x - (-8)x$. Use the distributive property.

$$\begin{aligned} 5x - (-8)x &= (5 * x) - (-8 * x) \\ &= (5 - (-8)) * x \\ &= (5 + 8) * x \\ &= 13 * x, \text{ or } 13x \end{aligned}$$

Check your answer by substituting several values for the variable.

Check: Substitute 5 for the variable.

$$\begin{aligned} 5x - (-8)x &= 13x \\ (5 * 5) - (-8 * 5) &= 13 * 5 \\ 25 - (-40) &= 65 \\ 65 &= 65 \end{aligned}$$

Check: Substitute 2 for the variable.

$$\begin{aligned} 5x - (-8)x &= 13x \\ (5 * 2) - (-8 * 2) &= 13 * 2 \\ 10 - (-16) &= 26 \\ 26 &= 26 \end{aligned}$$

If there are more than 2 like terms, you can add or subtract the terms in the order in which they occur and keep a running total.

Example 2: Simplify the expression $2n - 7n + 3n - 4n$.

$2n - 7n = -5n$

$-5n + 3n = -2n$

$-2n - 4n = -6n$

Therefore, $2n - 7n + 3n - 4n = -6n$.

Simplify each expression by rewriting it as a single term.

1. $6y + 13y$ _______________

2. $7g - 12g$ _______________

3. _______________ $5\frac{1}{2}x - 1\frac{1}{2}x$

4. $3c - (-5)c$ _______________

5. $5y - 3y + 11y$ _______________

6. $6g - 8g + 5g - 4g$ _______________

7. $n + n + n + n + n$ _______________

8. $n + 3n + 5n - 7n$ _______________

9. $2x + 4x - (-9)x$ _______________

10. $-7x + 2x + 3x$ _______________

Date Time

LESSON 9•3

Combining Like Terms *continued*

An expression such as $2y + 6 + 4y - 8 - 9y + (-3)$ is difficult to work with because it is made up of 6 different terms that are added and subtracted.

There are 2 sets of like terms in the expression. The terms $2y$, $4y$, and $9y$ are 1 set of like terms. The constant terms 6, 8, and (-3) are a second set of like terms. Each set of like terms can be combined into a single term. To simplify an expression that has more than one set of like terms, combine each set into a single term.

Example 3: Simplify $2y + 6 + 4y - 8 - 9y + (-3)$ by combining like terms.

Step 1 Combine the y terms. $2y + 4y - 9y = 6y - 9y = -3y$

Step 2 Combine the constant terms. $6 - 8 + (-3) = -2 + (-3) = -5$

Final result: $2y + 6 + 4y - 8 - 9y + (-3) = -3y + (-5) = -3y - 5$

Check: Substitute 2 for y in the original expression and the simplified expression.

$$2y + 6 + 4y - 8 - 9y + (-3) = -3y - 5$$
$$(2 * 2) + 6 + (4 * 2) - 8 - (9 * 2) + (-3) = (-3 * 2) - 5$$
$$4 + 6 + 8 - 8 - 18 + (-3) = -6 - 5$$
$$-11 = -11$$

Simplify each expression by combining like terms. Check each answer by substituting several values for the variable.

11. $4 + 7y + 20$ ________________

12. $5x - 3x + 8$ ________________

13. $5n + 6 - 8n - 2 - 3n$ ________________

14. $n + \pi + 2n - \frac{1}{2}\pi$ ________________

15. $-2.5x + 9 + 1.4x + 0.6$ ________________

16. $9d + 2a - (-6a) + 3d - 15d$ ________________

Date Time

LESSON 9·3 Estimating and Measuring in Millimeters

All measurements are approximations. The pencil shown below measures about 4 inches. A more precise measurement is about $3\frac{10}{16}$ inches. An even more precise measurement is about 93 millimeters.

The smaller the unit you use to measure an object, the more precise the measurement will be.

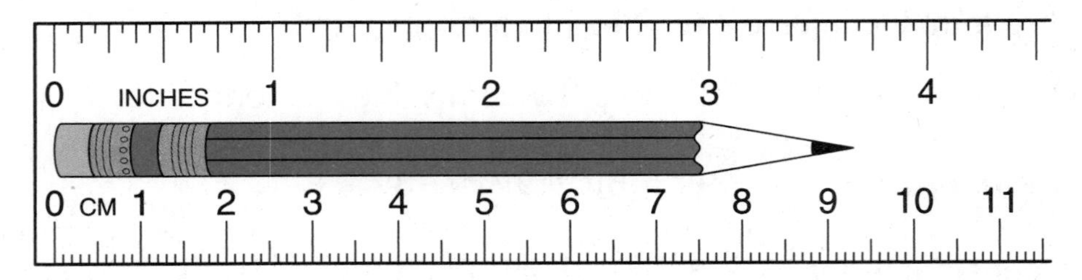

1. Explain why 93 millimeters is a more precise measurement than $3\frac{10}{16}$ inches for the length of the pencil.

____

____

Measure each line segment below to the nearest millimeter.

2. A ———— B Length of $\overline{AB}$ = ________ mm

3. D ———————————— E Length of $\overline{DE}$ = ________ mm

4. Measure $\overline{PQ}$. Then, in the space provided, draw a line segment that is $\frac{5}{8}$ the length of $\overline{PQ}$. Label the segment $\overline{RS}$.

   P ——————————— Q Length of $\overline{PQ}$ = ________ mm

   Length of $\overline{RS}$ = ________ mm

5. Measure $\overline{FG}$. Then, in the space provided, draw a line segment that is 125% the length of $\overline{FG}$. Label the segment $\overline{JK}$.

   F ———————————— G Length of $\overline{FG}$ = ________ mm

   Length of $\overline{JK}$ = ________ mm

Date Time

LESSON 9·3

Math Boxes

1. The area of the rectangle is 238 $units^2$.

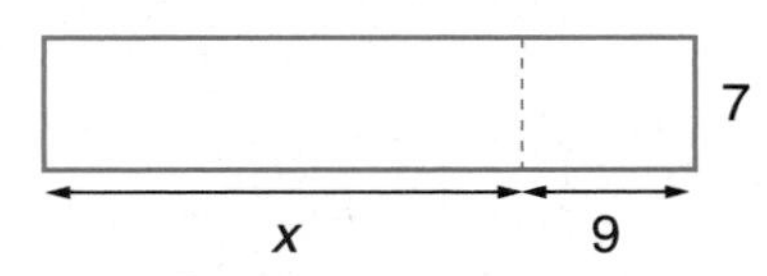

Write a number sentence to find the value of *x*.

Number sentence ____________________

Solve for *x*.

$x =$ ______ units

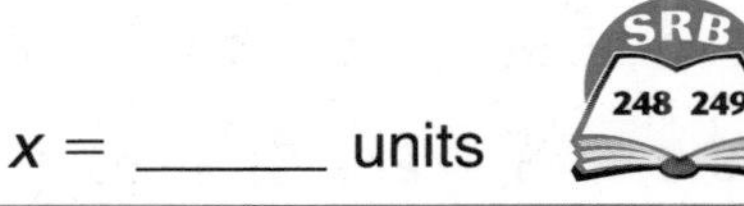

2. Solve.

a. $-100 = 12w - 4$

Solution __________

b. $-24 + p + 8 = 4$

Solution __________

c. $4j - 24 - 6j = 120$

Solution __________

3. Triangles *ABC* and *DEF* are similar.

a. Length of $\overline{AB}$ = ________ units

b. Length of $\overline{DF}$ = ________ units

c. $\frac{\text{Perimeter of triangle } ABC}{\text{Perimeter of triangle } DEF}$ = ________ units

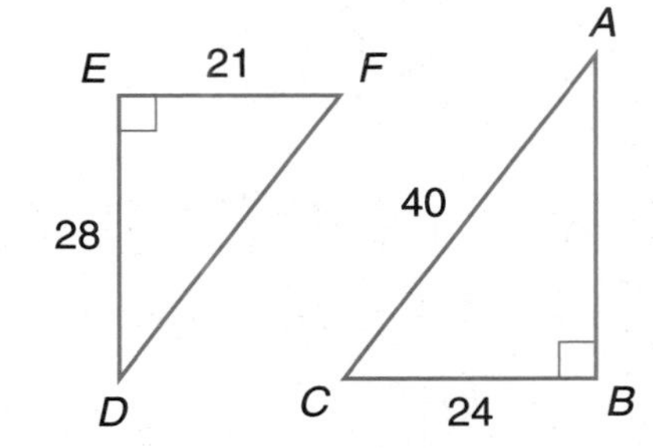

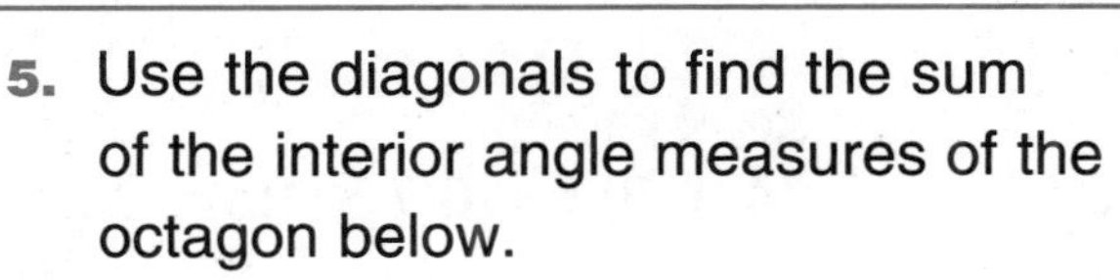

4. I am a regular polygon with all obtuse angles. I have the smallest number of sides of any polygon with obtuse angles. How many sides do I have?

____

Use your Geometry Template to draw this polygon below.

5. Use the diagonals to find the sum of the interior angle measures of the octagon below.

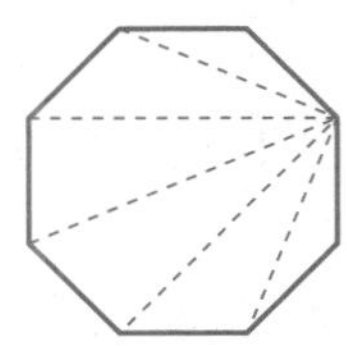

Sum of the interior angle measures = ____________ °

Date Time

LESSON 9•4

Simplifying Algebraic Expressions

Simplify the following expressions. First use the distributive property to remove the parentheses. Then combine like terms. Check the answer by substituting a value for the variable.

Example: Simplify $20 * (3 + 2x) + 30x$.

Step 1 Use the distributive property to remove the parentheses.

$$20 * (3 + 2x) + 30x = (20 * 3) + (20 * 2x) + 30x$$
$$= 60 + 40x + 30x$$
$$= 60 + 70x$$

Step 2 Simplify the expression by combining like terms.

Therefore, $20 * (3 + 2x) + 30x = 60 + 70x$

Check by substituting 5 for x.

$$20 * (3 + 2 * 5) + 30 * 5 = 60 + 70 * 5$$
$$20 * (3 + 10) + 150 = 60 + 350$$
$$20 * 13 + 150 = 410$$
$$260 + 150 = 410$$
$$410 = 410$$

1. $7 + (5 - 3) * x + 1$ ________________

2. $2(g - 1) + 1 - 5g$ ________________

3. $\frac{1}{2}(2m + 1) + \frac{1}{2}$ ________________

4. $n + 2n + 3n + (4 + 5)n + 6(7 + 2n)$ ________________

Try This

5. $6(p - 7) - 5p + 15 + (3p + 2)4$ ________________

6. $12.4(2f - 5) - 11.6(3f - 2) + 3.4(0.5f + 2)$ ________________

Date Time

LESSON 9·4

Simplifying Equations

Simplify both sides of the following equations. Do NOT solve them.

Example: $2b + 5 + 3b = 8 - b + 21$

$(2b + 3b) + 5 = 8 - b + 21$

$5b + 5 = 8 - b + 21$

$5b + 5 = -b + 21 + 8$

$5b + 5 = -b + 29$

1. $5h + 13h = 20 - 2$

2. $2 + x + 2x + 4 = x + 16$

3. $2(y + 2) = 4(y + 3)$

4. $(4 - 1)m - m = (m - 1) * 4$

5. $4y + 6 = 8(1 + y)$

6. $5(x + 3) - 2x = 35 + x$

7. $3 * (3.2 - 2c) = 4.6 + 4c$

8. $4(2z - 5) = z + 1$

LESSON 9•4 Math Boxes

1. Simplify each expression.

   a. $8(2y - 3) - 6y$ ______

   b. $18 - 4(8m + 5)$ ______

   c. $-(6w - 1) + 3(w - 4)$

   ______

2. Circle the expressions that represent the area of the shaded rectangle.

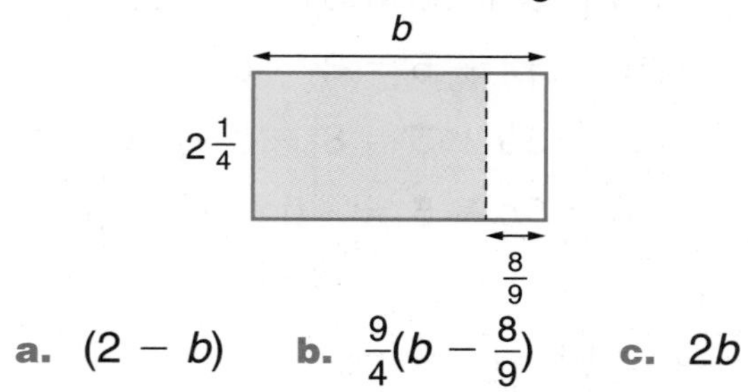

   a. $(2 - b)$ b. $\frac{9}{4}(b - \frac{8}{9})$ c. $2b$

   d. $\frac{9}{4}b$ e. $2\frac{1}{4}b - 2$ f. $2 - 2\frac{1}{4}b$

3. Find the number.

   a. 75% of what number is 96? ______

   b. 40% of what number is 30? ______

   c. $33\frac{1}{3}$% of what number is 48? ______

   d. 10% of what number is 11? ______

   e. $12\frac{1}{2}$% of what number is 50? ______

4. Write >, <, or =.

   a. $12 - (-3)$ ______ $\frac{7}{8} \div \frac{1}{20}$

   b. $5^2 + 3^3$ ______ $5\frac{20}{4} + 10\frac{50}{10}$

   c. $3\frac{6}{7} + 2\frac{3}{5}$ ______ $\frac{100}{18}$

   d. $0.48 * 2.5$ ______ $3 * 0.26$

   e. $-4 * -8$ ______ $-(2^5)$

5. Plot and label points on the coordinate grid as directed.

   a. Plot (−4, −3). Label it *L*.

   b. Plot (4, −1). Label it *M*.

   c. Draw line segment *LM*.

   d. Name the coordinates of the midpoint of $\overline{LM}$.

   (______, ______)

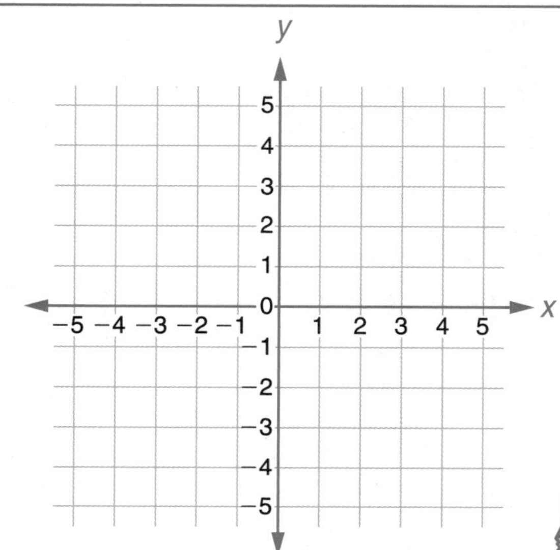

SRB 234

Date Time

LESSON 9•5

Number Stories and the Distributive Property

Solve each problem mentally. Then record the number model you used.

1. A carton of milk costs $0.60. John bought 3 cartons of milk one day and 4 cartons the next day.

   How much did he spend in all? ____________________

   Number model ____

2. During a typical week, Karen runs 16 miles and Jacob runs 14 miles.

   About how many miles in all do Karen and Jacob run in 8 weeks? ____________________

   Number model ____

3. Mark bought 6 CDs that cost $12 each. He returned 2 of them.

   How much did he spend in all? ____________________

   Number model ____

4. Max collects stamps. He had 9 envelopes, each containing 25 stamps. He sold 3 envelopes to another collector.

   How many stamps did he have left? ____________________

   Number model ____

5. Jean is sending party invitations to her friends. She has 8 boxes with 12 invitations in each box. She has already mailed 5 boxes of invitations.

   How many invitations are left? ____________________

   Number model ____

Date Time

LESSON 9•5

Simplifying and Solving Equations

Simplify each equation. Then solve it. Record the operations you used for each step.

1. $6y - 2y = 40$

Solution ____________

2. $5p + 28 = 88 - p$

Solution ____________

3. $8d - 3d = 65$

Solution ____________

4. $12e - 19 = 7 - e$

Solution ____________

5. $3n + \frac{1}{2}n = 42$

Solution ____________

6. $3m - 1 + m + 6 = 2 - 9$

Solution ____________

7. $3(1 + 2y) = y + 2y + 4y$

Solution ____________

8. $8 - 12x = 6 * (1 + x)$

Solution ____________

9. $-4.8 + b + 0.6b = 1.8 + 3.6b$

Solution ____________

10. $4t - 5 = t + 7$

Solution ____________

Date ____________________ Time ____________________

LESSON 9•5 Simplifying and Solving Equations *continued*

11. $8v - 25 = v + 80$

Solution ____________________

12. $3z + 6z = 60 - z$

Solution ____________________

13. $g + 3g + 32 = 27 + 5g + 2$

Solution ____________________

14. $16 + 3s - 2s = 24 + 2s - 20$

Solution ____________________

15. Are the following 2 equations equivalent? __________

$5y + 3 = -6y + 4 + 12y$ $\quad$ $5y + 3 = -6y + 4(1 + 3y)$

Explain your answer. ____

____

____

16. Are the following 2 equations equivalent? __________

$5(f - 2) + 6 = 16$ $\quad$ $f - 1 = 3$

Explain your answer. ____

____

____

Try This

17. Solve $\frac{2z + 4}{5} = z - 1$

(*Hint:* Multiply both sides by 5.)

Solution ____________________

Date Time

LESSON 9•5 Math Boxes

1. The area of the shaded part of the rectangle is 20 units^2.

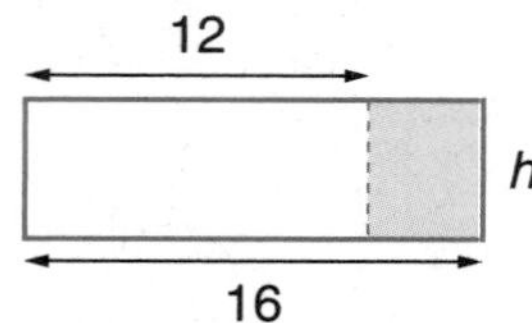

Write a number sentence to find the value of h.

Number sentence: ____________________

Solve for h.

h = ______ units

2. Solve.

a. $\frac{1}{3}f - 6 = -8$

Solution __________

b. $-30 = b - 6 + 11b$

Solution __________

c. $4g + 4 = 2g + 36$

Solution __________

3. Triangles THG and TIN are similar.

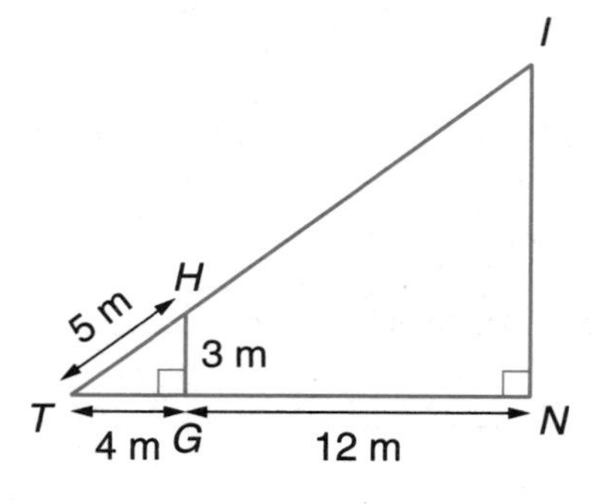

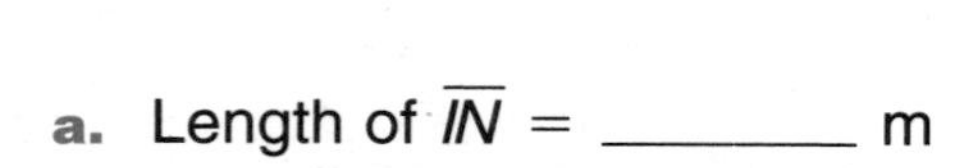

a. Length of $\overline{IN}$ = ________ m

b. Length of $\overline{HI}$ = ________ m

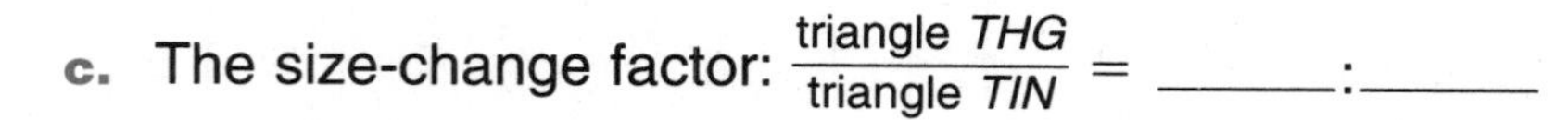

c. The size-change factor: $\frac{\text{triangle } THG}{\text{triangle } TIN}$ = ______:______

4. I am a quadrangle with 2 pairs of congruent adjacent sides. One of my diagonals is also my only line of symmetry. How many sides do I have?

____

Use your Geometry Template to draw this polygon in the space provided at the right.

5. The polygon below is a regular polygon. Find the measure of angle X without using a protractor.

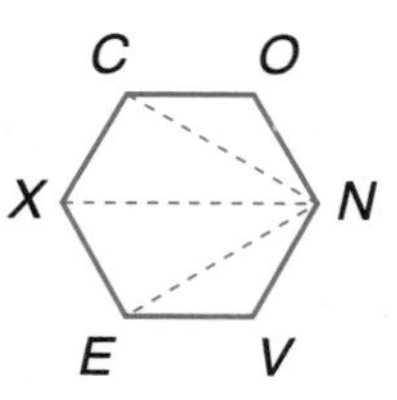

m∠X = ________°

Date Time

LESSON 9•6

Mobile Problems

The mobile shown in each problem is in balance.
The **fulcrum** of the mobile at the right is the center point of the rod.
A mobile will balance if $W * D = w * d$.

Write and solve an equation to answer each question.

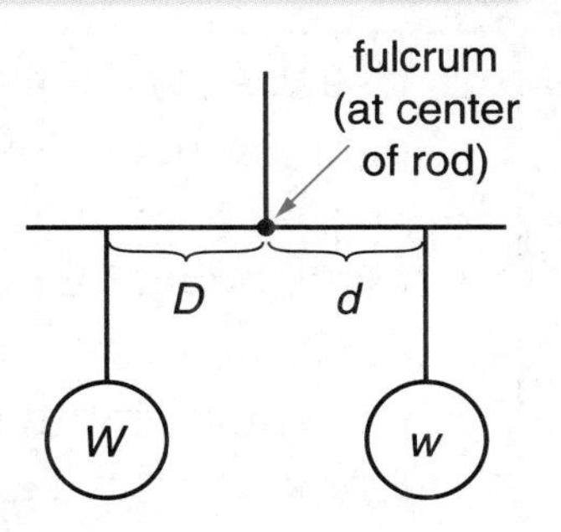

1. What is the distance from the fulcrum to the object on the right of the fulcrum?

   $W =$ _______ $D =$ _______ $w =$ _______ $d =$ _______

   Equation ____________________ Solution ____________

   Distance _______ units

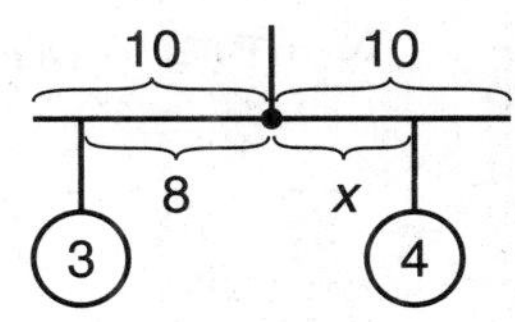

2. What is the weight of the object on the left of the fulcrum?

   $W =$ _______ $D =$ _______ $w =$ _______ $d =$ _______

   Equation ____________________ Solution ____________

   Weight _______ units

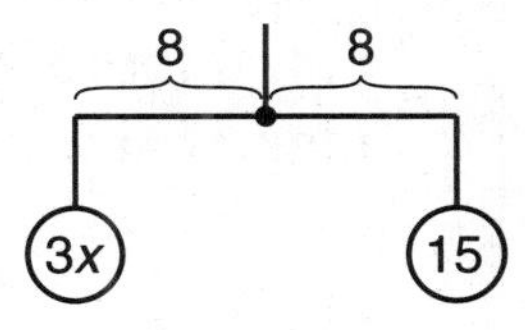

3. What is the distance from the fulcrum to each of the objects?

   $W =$ _______ $D =$ _______ $w =$ _______ $d =$ ____________

   Equation ____________________ Solution ____________

   Distance on the left of the fulcrum _______ units

   Distance on the right of the fulcrum _______ units

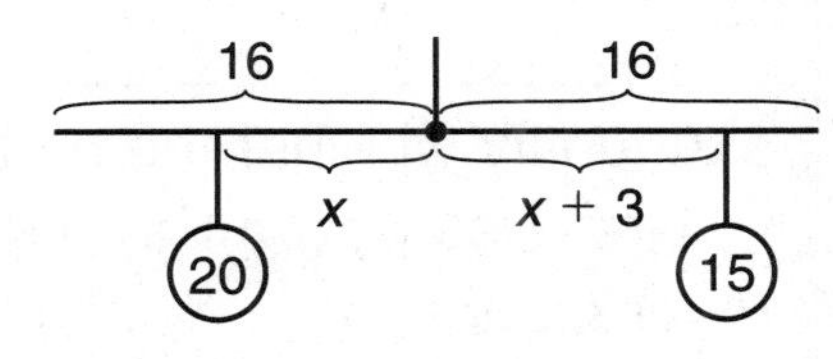

4. What is the weight of each object?

   $W =$ ____________ $D =$ _______ $w =$ ____________ $d =$ _______

   Equation ____________________ Solution ____________

   Weight of the object on the left of the fulcrum _________ units

   Weight of the object on the right of the fulcrum _________ units

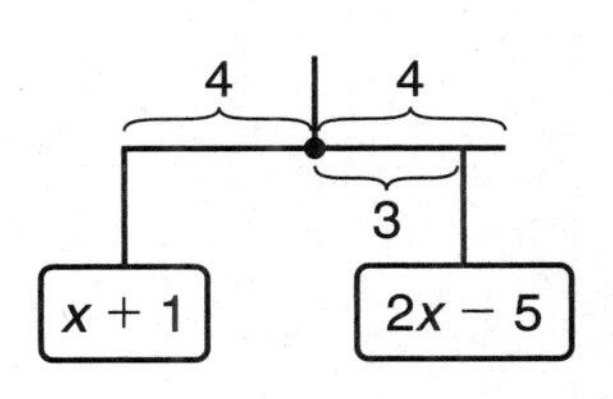

Date Time

LESSON 9•6

Math Boxes

1. Use the distributive property to write a number model for the problem. Then solve.

Tyrell bought 2 pairs of jeans for $34.99 each and 2 T-shirts for $19.99 each. How much more did he spend on jeans than on T-shirts?

Number model

____

Solution ________

2. Simplify each expression by combining like terms.

a. $9x + 12 - x$ ____________

b. $h - 14 - 2h + 8$ ____________

c. $8d - d - 5m$ ____________

d. $4w + 4t - 3w - 9t$ ____________

3. The ratio of facedown to faceup cards is 5:4. If there are 72 cards altogether, how many cards are faceup?

________ cards

4. Draw the line(s) of symmetry for each figure below.

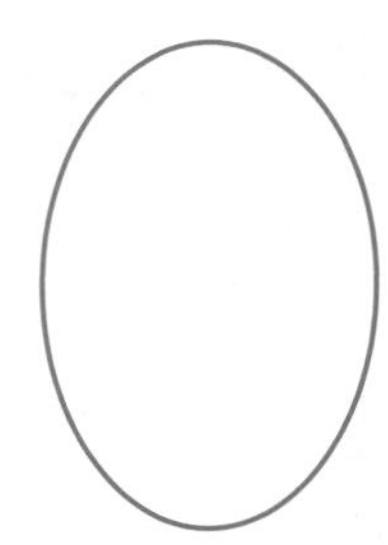

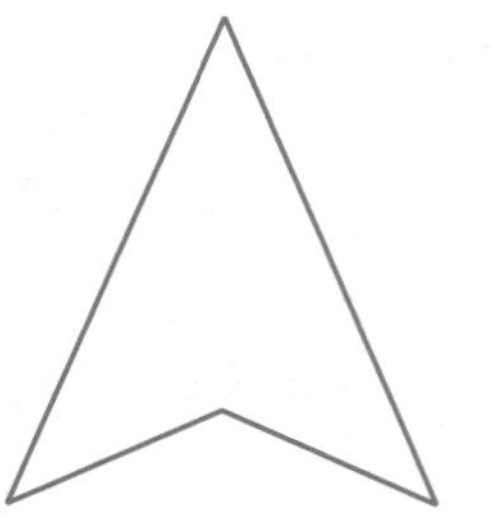

SRB 182

5. Identify whether the preimage (1) and image (2) are related by a translation, a reflection, or a rotation.

Write your answer on the line below.

____

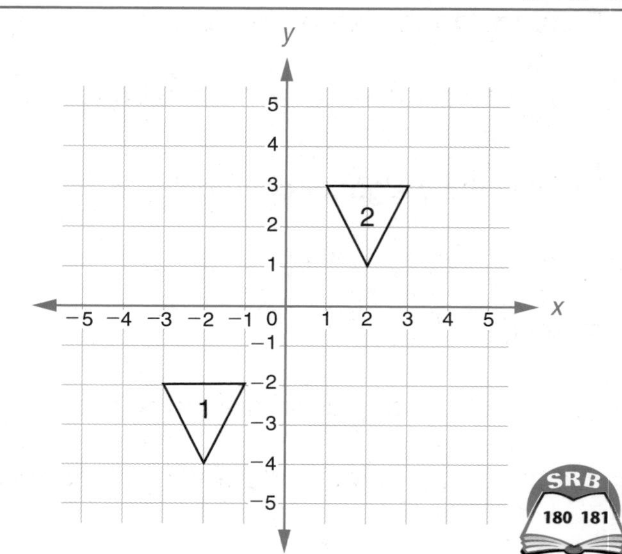

SRB 180 181

Date Time

LESSON 9•7

A Picnic Budget Spreadsheet

The following spreadsheet gives budget information for a class picnic.

Class Picnic ($)

	A	B	C	D
1		budget for class picnic		
2				
3	quantity	food items	unit price	cost
4	6	packages of hamburgers	2.79	16.74
5	5	packages of hamburger buns	1.29	6.45
6	3	bags of potato chips	3.12	9.36
7	3	quarts of macaroni salad	4.50	13.50
8	4	bottles of soft drinks	1.69	6.76
9			subtotal	52.81
10			8% tax	4.23
11			total	57.04

1. What information is shown in Row 8?

____

2. What information is shown in column A—labels, numbers, or formulas? _______________

3. Cell D6 holds the following formula: = A6 ∗ C6.

   a. What formula is stored in cell D4? _______________

   b. What formula is stored in cell D8? _______________

4. Circle the formula stored in cell D9.

   = C4 + C5 + C6 + C7 + C8 = D4 + D5 + D6 + D7 + D8

5. a. What does the formula stored in cell D10 calculate? _______________

   b. Circle the formula stored in cell D10.

   = 0.08 ∗ C9 = 0.08 ∗ D9 = 8 ∗ D9

6. a. What does the formula stored in cell D11 calculate? _______________

   b. Write the formula stored in cell D11. _______________

7. a. Which spreadsheet cells would change if you increased the number of bags of potato chips to 4? _______________

   b. Calculate the number that would be shown in each of these cells.

____

Date Time

LESSON 9•7

Math Boxes

1. Use the distributive property to remove the parentheses. Then combine like terms.

a. $-5 + 5(x + 4)$ ____________

b. $3m + 2(5m - 7)$ ____________

c. $4(6t + 9) - 10t$ ____________

d. $8k - 6(3 - 2k)$ ____________

2. Solve each equation.

a. $8b - 10 = 2b + 14$

$b =$ ____________

b. $4 - 6w = 3(\frac{1}{2} + \frac{w}{2})$

$w =$ ____________

3. Circle the equation that describes the relationship between the numbers in the table at the right.

A. $(x - 9)5 = y$

B. $\frac{x - 9}{5} = y$

C. $(y + 5)9 = x$

D. $5(y + 5) = x$

x	y
10	$\frac{1}{5}$
14	1
19	2
49	8

4. Without using a protractor, find the measure of each numbered angle below. Write each measure on the drawing.

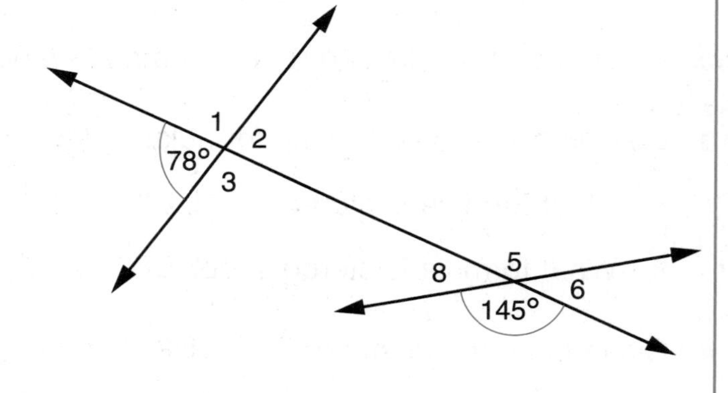

5. Using a compass, draw a circle that has a circumference of 15.7 cm. Label the diameter of your circle.

Use 3.14 for π.

Date Time

LESSON 9•8 Area Formulas

Calculate the area of each figure below. A summary of useful area formulas appears on page 377 of the *Student Reference Book.*

Measure dimensions to the nearest tenth of a centimeter. Record the dimensions next to each figure. You might need to draw and measure 1 or 2 line segments on a figure. Round your answers to the nearest square centimeter.

1.

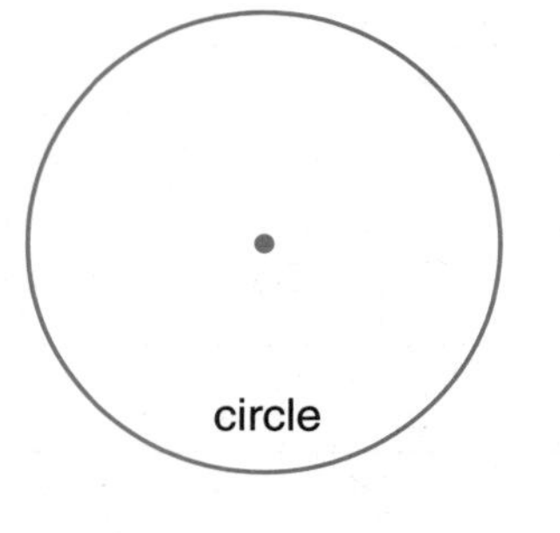

Area formula ______________

Area ______________ (unit)

2.

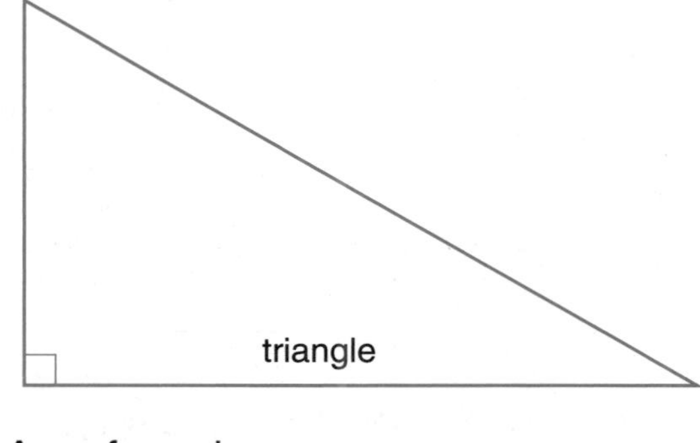

Area formula ______________

Area ______________ (unit)

3.

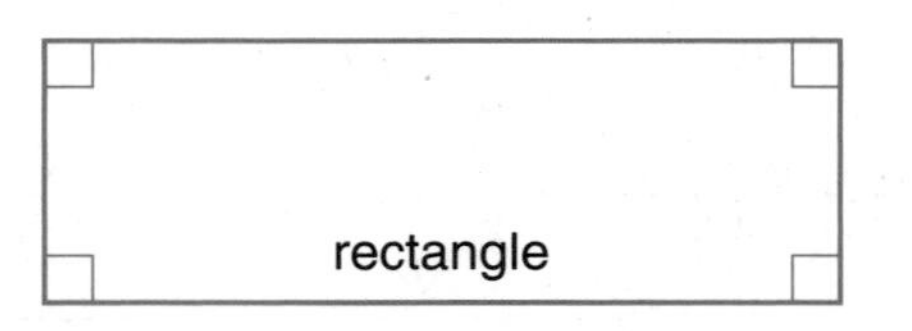

Area formula ______________

Area ______________ (unit)

4.

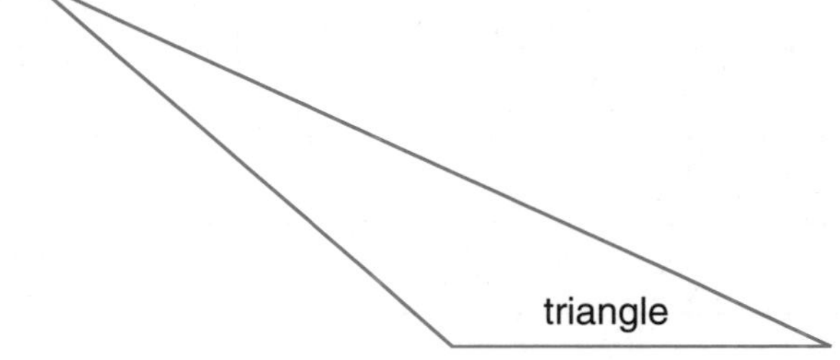

Area formula ______________

Area ______________ (unit)

5.

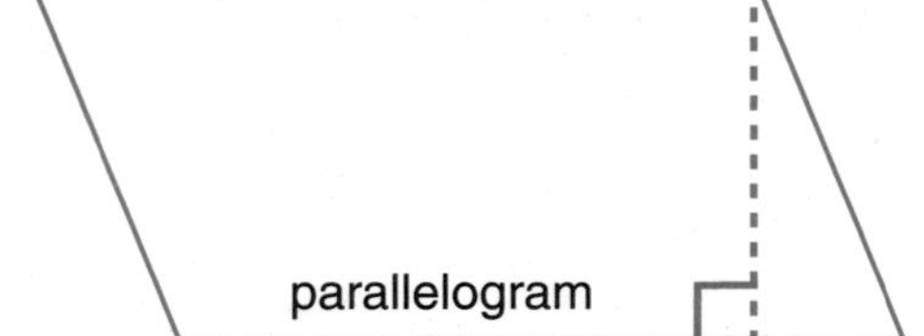

Area formula ______________

Area ______________ (unit)

Try This

6.

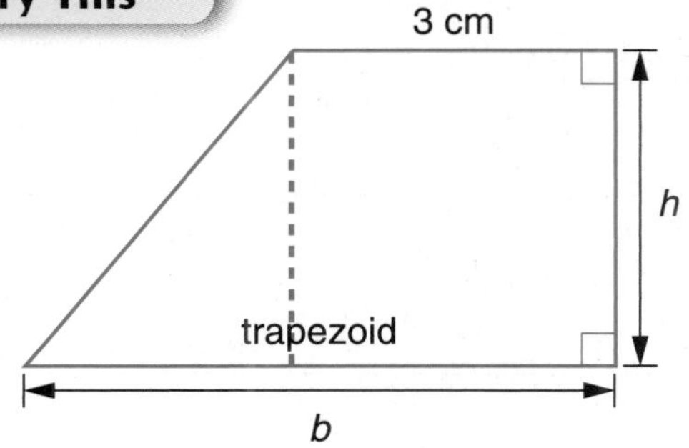

Area formula ______________

Area ______________ (unit)

Date Time

LESSON 9•8

Perimeter, Circumference, and Area

Solve each problem. Explain your answers.

1. Rectangle *PERK* has a perimeter of 40 feet.

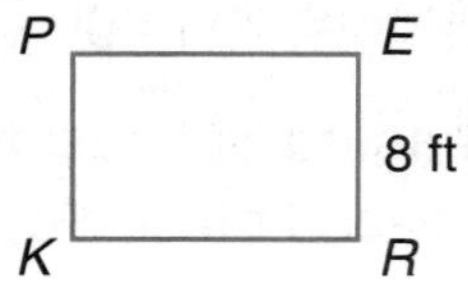

Length of side *PE* ________ (unit)

Area of rectangle *PERK* ________ (unit)

____

____

____

____

2. The area of triangle *BAC* is 300 meters^2.
What is the length of side *AB?*

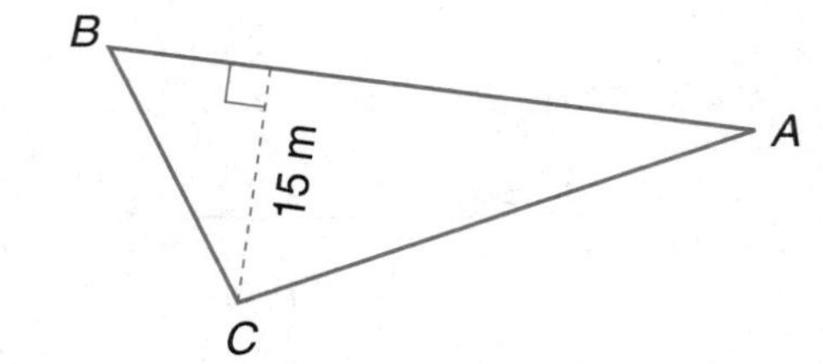

Length of side *AB* ________ (unit)

____

____

____

____

3. The area of parallelogram *LMNK* is 72 square inches.

The length of side *LX* is 6 inches, and the length of side *KY* is 3 inches.

What is the length of $\overline{LY}$?

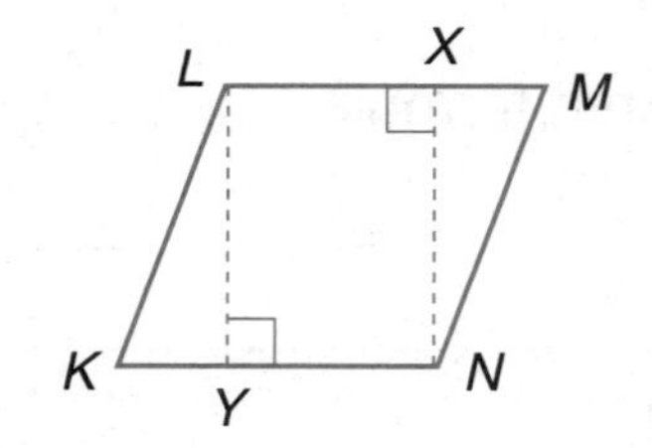

Length of $\overline{LY}$ ________ (unit)

____

____

____

____

Date Time

LESSON 9•8

Perimeter, Circumference, and Area *cont.*

4. The area of triangle *ACE* is 42 square yards. What is the area of rectangle *BCDE?*

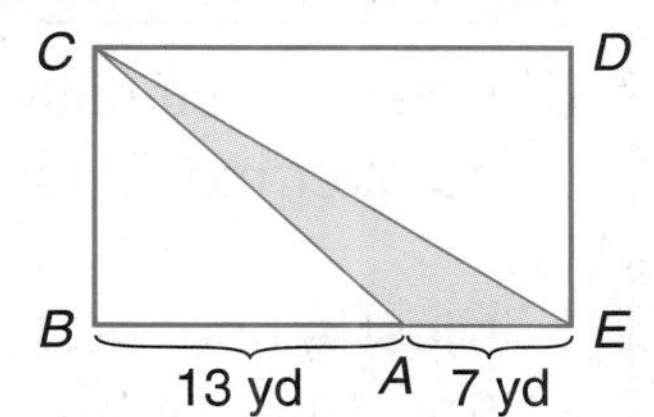

Area of rectangle *BCDE* __________

For Problems 5 and 6, use 3.14 for π.

5. To the nearest percent, about what percent of the area of the square is covered by the area of the circle?

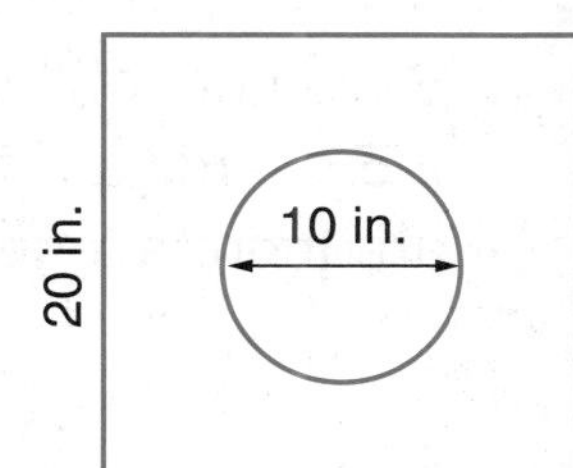

Answer __________

6. Which path is longer: once around the figure 8—from *A* to *B* to *C* to *B* and back to *A*—or once around the large circle?

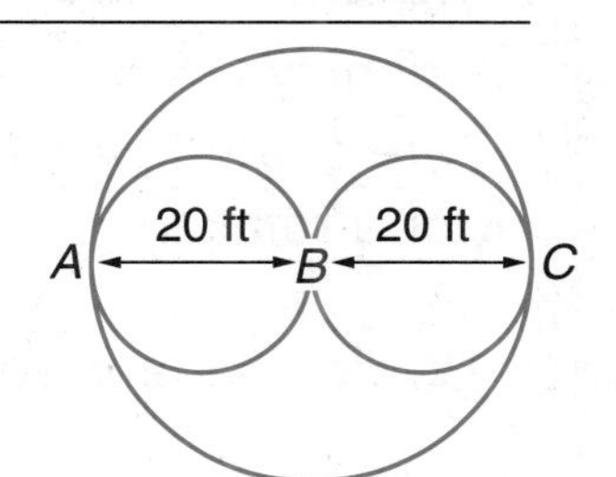

Date Time

LESSON 9•8

Math Boxes

1. Use the distributive property to write a number model for the problem. Then solve.

   Niesha bought 3 sets of screwdrivers at \$19.49 each and 3 boxes of screws at \$4.98 each. What was the total cost of Niesha's purchase?

   Number model

   ____

   Solution ________

2. Simplify each expression by combining like terms.

   a. $3m + 4p + 6m + 9$ ________

   b. $2k + 7n - 9n - 4k$ ________

   c. $\frac{3}{2}w - (-\frac{4}{5}) + \frac{5}{2}w$ ________

   d. $-6x - 6y - 6x - 7y$ ________

3. Seven out of 9 cards are faceup. If 56 cards are faceup, how many cards are there altogether?

   ________ cards

4. Draw the line(s) of symmetry for each figure below.

5. Identify whether the preimage (1) and image (2) are related by a translation, a reflection, or a rotation.

   Write your answer on the line below.

   ____

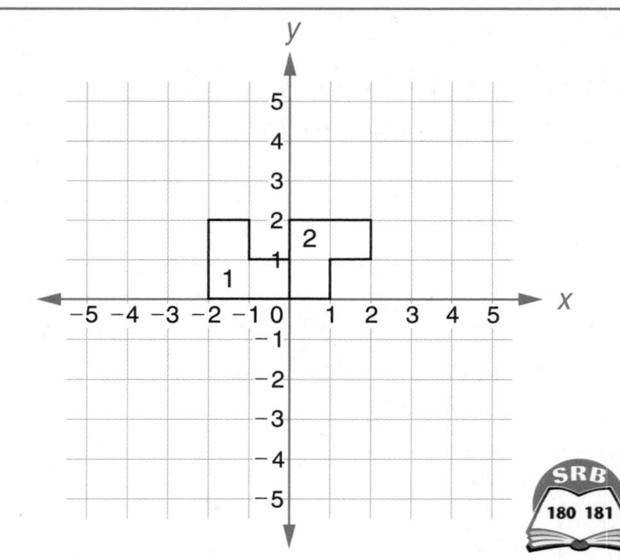

Date ______________________ Time ______________________

LESSON 9•9 Data Review

1. Below are the scores for a spelling test in Ms. Jenning's sixth-grade class:

72%	96%	88%	96%	80%	68%	44%
76%	96%	68%	56%	76%	96%	92%
80%	88%	68%	56%	100%	100%	88%
68%	96%	92%	96%	76%	80%	88%

a. Make a stem-and-leaf plot of the scores.

b. Find the following landmarks:

maximum ____________ median ____________

mode(s) ____________ minimum ____________

2. First Bank and Trust raised the interest rate on savings accounts 4 times in 1 year. To the right is a step graph of the interest rates for the year. Use the graph to answer the questions.

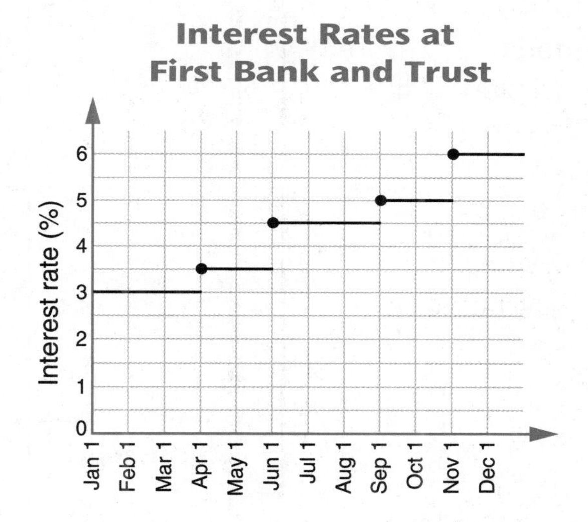

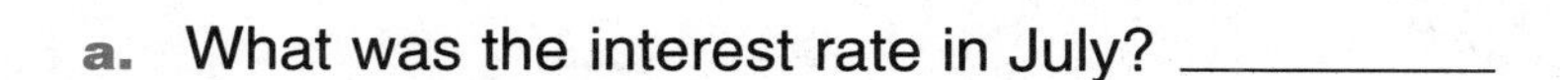

a. What was the interest rate in July? ____________

b. For how many months did the interest rate stay at 4.5%? ____________

c. By how much did the interest rate increase from February to October? ____________

3. Which graph below most likely displays the number of cellular phone subscribers (in millions)? Graph ______

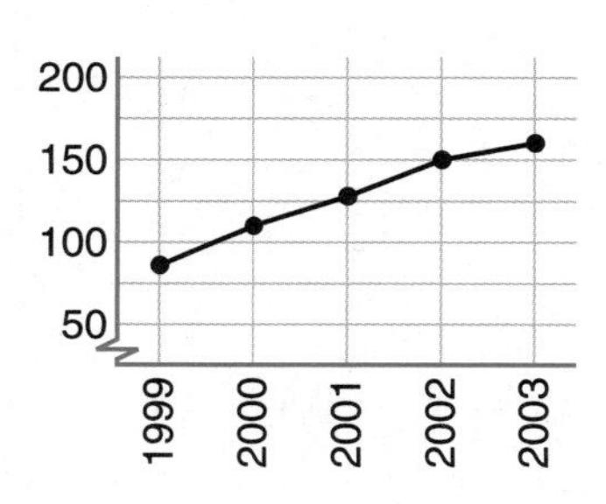

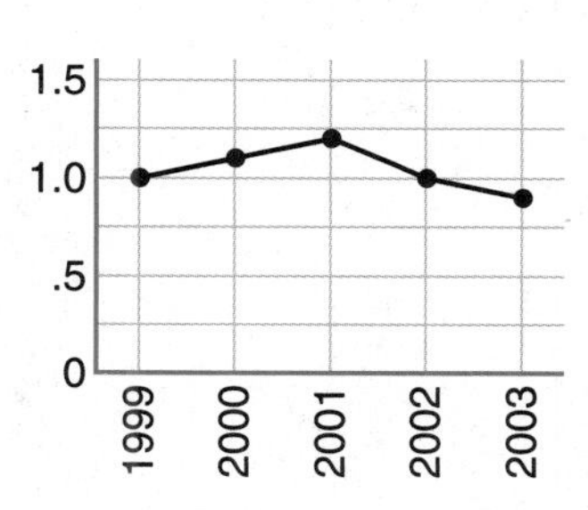

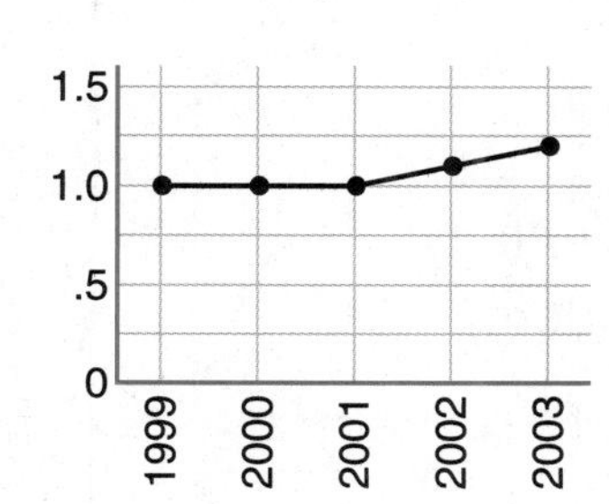

LESSON 9•9 Calculating the Volume of the Human Body

An average adult human male is about 69 inches (175 centimeters) tall and weighs about 170 pounds (77 kilograms). The drawings below show how a man's body can be approximated by 7 cylinders, 1 rectangular prism, and 1 sphere.

The drawings use the scale 1 mm:1 cm. This means that every length of 1 millimeter in the drawing represents 1 centimeter of actual body length. The drawing below is 175 millimeters high. Therefore, it represents a male who is 175 centimeters tall.

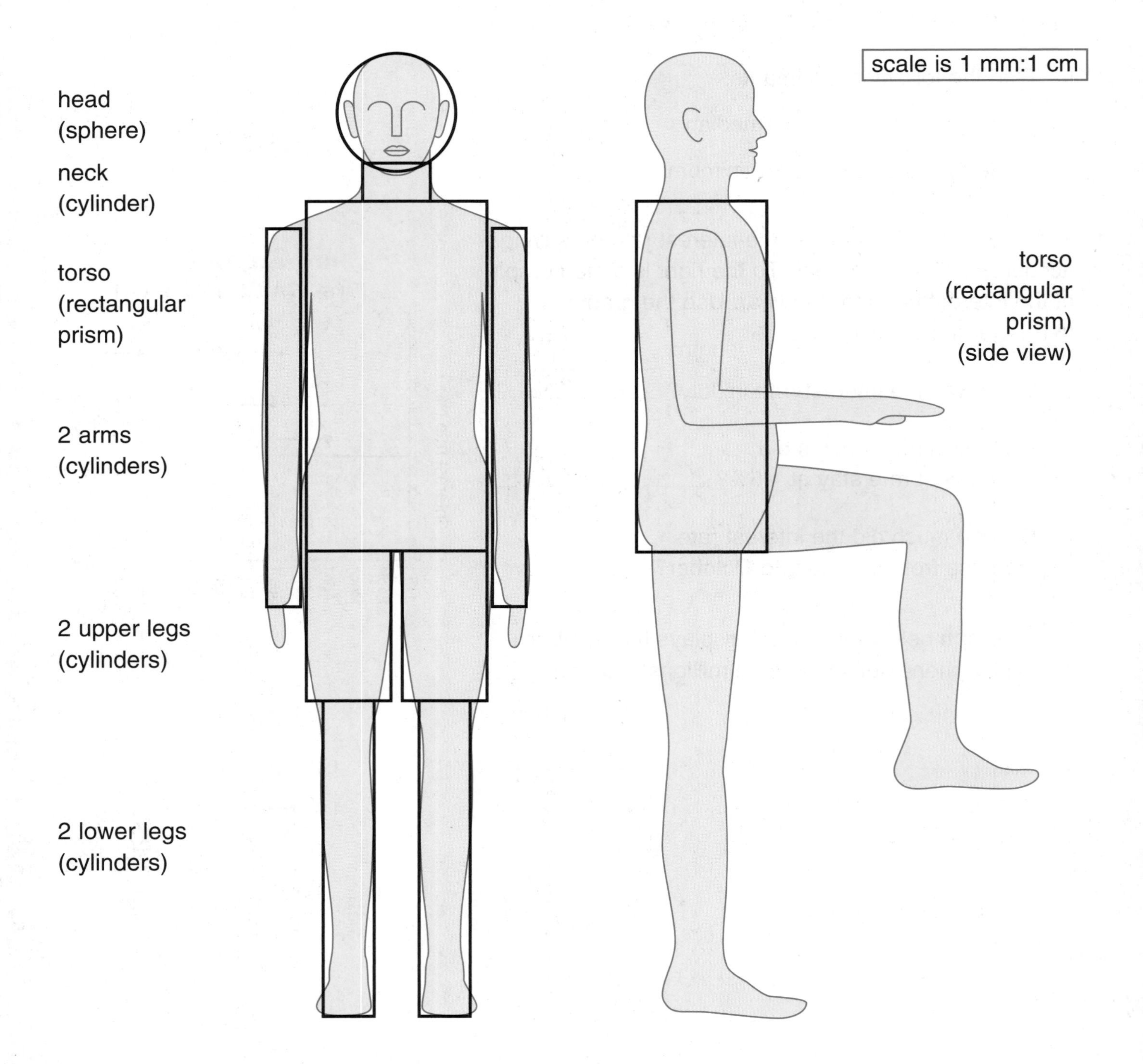

Date Time

Calculating the Volume of the Human Body *cont.*

1. a. Use a centimeter ruler to estimate the measures of each geometric figure shown on page 350. Record your estimates on the drawings and in the table below. Be sure to record the actual body dimensions. For example, if you measure the length of an arm as 72 millimeters, record this as 72 centimeters because the scale of the drawing is 1 mm:1 cm.

b. Calculate the volume of each body part and record it in the table. You will find a summary of useful volume formulas on page 378 in your *Student Reference Book.*

For the arm, upper leg, and lower leg, multiply the volume by 2. Add to find the total volume of an average adult male's body. Your answer will be in cubic centimeters.

Body Part and Shape	Actual Body Dimensions (cm)	Volume (Round to the nearest 1,000 cm^3.)
Head (sphere)	radius:	∗ 1 =
Neck (cylinder)	radius: height:	∗ 1 =
Torso (rectangular prism)	length: width: height:	∗ 1 =
Arm (cylinder)	radius: height:	∗ 2 =
Upper leg (cylinder)	radius: height:	∗ 2 =
Lower leg (cylinder)	radius: height:	∗ 2 =
	Total Volume:	About

2. One liter is equal to 1,000 cubic centimeters. Use this fact to complete the following statement: I estimate that the total volume of an average adult male's body is about ________ liters.

3. A men's size 7 regulation basketball has a radius of about 12 cm.

a. Find the volume of this type of basketball, rounded to the nearest 1,000 cm^3. ____________ (unit)

b. Fill in the blank. The volume of a men's size 7 regulation basketball is about ________% the volume of an average adult male's head.

Date Time

LESSON 9•9

Math Boxes

1. Use the distributive property to remove parentheses. Then combine like terms.

a. $3(t + 3) + 4t$ ____________

b. $19 - 4(5y + 1) - 4y$ ____________

c. $10 + 7m - 2(3m + 5)$ ____________

d. $-7(2p - 1) + 3(8 - p)$ ____________

2. Solve each equation.

a. $-2(g + 6) = g + 3 + 2g$

$g =$ ____________

b. $2(2x + \frac{1}{2}) = 3(x - \frac{2}{3})$

$x =$ ____________

3. Circle the equation that describes the relationship between the numbers in the table at the right.

A. $(x * 4) - 3 = y$

B. $(4 * x) + 3 = y$

C. $(y * 5) - 3 = x$

D. $(4 * y) + 3 = x$

x	*y*
$\frac{1}{4}$	−2
$\frac{1}{2}$	−1
4	13
10	37

4. Without using a protractor, find the measure of each numbered angle below. Write each measure on the drawing. Lines *a* and *b* are parallel.

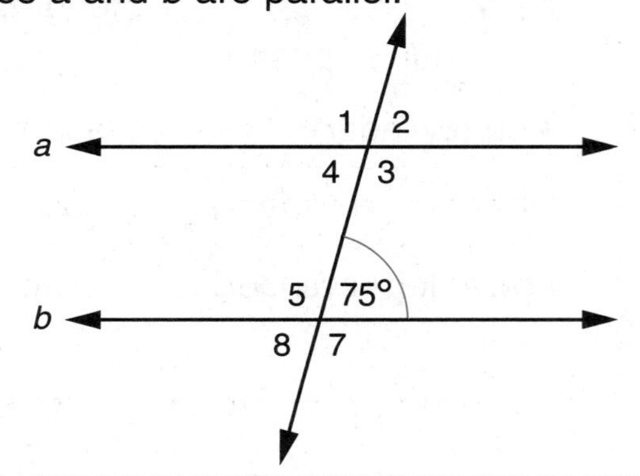

5. a. Using a compass, draw 2 concentric circles in the space at the right. The radius of the small circle is 1.5 cm. The radius of the large circle is 2 cm.

b. What is the area of the ring between the 2 circles? Use 3.14 for π.

About ____________ cm^2

Date Time

LESSON 9•10

Math Boxes

1. Calculate the area of the parallelogram.

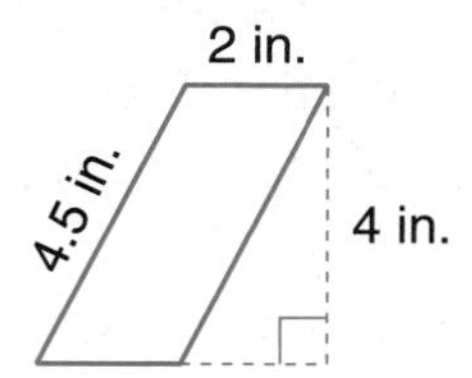

Area = ______
(unit)

2. Solve.

a. $0.25t = 645$ ______

b. $d * 10^2 = 420.5$ ______

c. $\frac{1}{8}f = \frac{3}{16}$ ______

d. $\sqrt{h} = 20$ ______

3. The table at the right shows about how much a person weighing 100 pounds on Earth would weigh on each of the planets in our solar system.

a. On which planet would a person weigh about $\frac{1}{6}$ as much as on Mercury?

______

b. On which planet would a person weigh about $2\frac{1}{2}$ times what the person weighs on Earth?

______

Planet	Weight (lb)
Mercury	37
Venus	88
Earth	100
Mars	38
Jupiter	264
Saturn	115
Uranus	93
Neptune	122
Pluto	6

4. a. Draw and label an obtuse angle *CAT*. Measure it.

$\angle CAT$ measures ______°.

b. Draw and label a reflex angle *NOD*. Measure it.

$\angle NOD$ measures ______°.

Date Time

LESSON 9•10 Solving Equations by Trial and Error

If you substitute a number for the variable in an equation and the result is a true number sentence, then that number is a solution of the equation. One way to solve an equation is to try several **test numbers** until you find the solution. Each test number can help you close in on an exact solution. Using this **trial-and-error method** for solving equations, you may not find an exact solution, but you can come very close.

Example: Find a solution of the equation $\frac{1}{x} + x = 4$ by trial and error. If you cannot find an exact solution, try to find a number that is very close to an exact solution.

The table shows the results of substituting several test numbers for x.

x	$\frac{1}{x}$	$\frac{1}{x} + x$	Compare $(\frac{1}{x} + x)$ to 4.
1	1	2	Less than 4
2	0.5	2.5	Still less than 4, but closer
3	0.3	3.3	Less than 4, but even closer
4	0.25	4.25	Greater than 4

These results suggest that we try testing numbers for x that are between 3 and 4.

x	$\frac{1}{x}$	$\frac{1}{x} + x$	Compare $(\frac{1}{x} + x)$ to 4.
3.9	0.256...	4.156...	> 4
3.6	$0.2\overline{7}$	$3.8\overline{7}$	< 4

Now it's your turn. Try other test numbers. See how close you can get to 4 for the value of $\frac{1}{x} + x$.

x	$\frac{1}{x}$	$\frac{1}{x} + x$	Compare $(\frac{1}{x} + x)$ to 4.

My closest solution: ____________

Date Time

LESSON 9•10 Solving Equations by Trial and Error *cont.*

Find numbers that are closest to the solutions of the equations. Use the suggested test numbers to get started. Round approximate solutions to the nearest thousandth.

1. Equation: $\sqrt{y} + y = 10$

y	$\sqrt{y}$	$\sqrt{y} + y$	Compare $(\sqrt{y} + y)$ to 10.
0	0	0	< 10
5	2.236	7.236	
9	3		

My closest solution: ____________

2. Equation: $x^2 - 3x = 8$

x	x^2	$3x$	$x^2 - 3x$	Compare $(x^2 - 3x)$ to 8.
4				
6				
5				

My closest solution: ____________

Date Time

LESSON 9•11

Using Formulas to Solve Problems

To solve a problem using a formula, you can substitute the known quantities for variables in the formula and solve the resulting equation.

Example: A formula for converting between Celsius and Fahrenheit temperatures is $F = 1.8C + 32$, where C represents the Celsius temperature and F represents the Fahrenheit temperature.

- Use the formula to convert 30°C to degrees Fahrenheit.

	$F = 1.8C + 32$
Substitute 30 for C in the formula.	$F = (1.8 * 30) + 32$
Solve the equation.	$F = 86$
Answer:	30°C = 86°F

- Use the formula to convert 50°F to degrees Celsius.

	$F = 1.8C + 32$
Substitute 50 for F in the formula.	$50 = (1.8 * C) + 32$
Solve the equation.	$10 = C$
Answer:	50°F = 10°C

1. The formula $W = 570a - 850$ expresses the relationship between the average number of words children know and their ages (for ages 2 to 8). The variable W represents the number of words known, and a represents age in years.

a. About how many words might a $3\frac{1}{4}$-year-old child know? __________

b. About how old might a child be who knows about 1,700 words? __________

2. A bowler whose average score is less than 200 is given a handicap. The **handicap** is a number of points added to a bowler's score for each game. A common handicap formula is $H = 0.8 * (200 - a)$, where H is the handicap and a is the average score.

a. What is the handicap of a bowler whose average score is 160? __________

b. What is the average score of a bowler whose handicap is 68? __________

3. An adult human female's height can be estimated from the length of her tibia (shinbone) by using the formula $H = 2.4 * t + 75$, where H is the height in centimeters and t is the length of the tibia in centimeters.

a. Estimate the height of a female whose tibia is 31 centimeters long. __________

b. Estimate the length of a female's tibia if she is 175 centimeters tall. __________

Date ______________________ Time ______________________

LESSON 9•11

Volume Problems

Solve each problem. You may need to look up formulas in your *Student Reference Book.* Use substitution to check your answers.

1. The volume of the desk drawer shown at the right is 1,365 $in.^3$
Find the depth (d) of the drawer.

6.5 in.
10 in.
d

Formula ______________________

Substitute ______________________

Solve ______________________

____

Depth of drawer = ______________

2. The cylindrical can holds about 4 liters (4 liters = 4,000 cm^3).
Find the height (h) of the can to the nearest centimeter. Use 3.14 for π.

8 cm
h

Formula ______________________

Substitute ______________________

Solve ______________________

____

The can's height is about ______________.

3. A soccer ball has a 9-inch diameter.

9 in.

a. What is the shape of the smallest box that will hold the ball? ______________

b. What are the dimensions of this box? ______________________

c. Compare the volume of the box to the volume of the ball. Is the volume of the box more or less than twice the volume of the ball? ______________
(*Reminder:* A formula for finding the volume of a sphere is $V = \frac{4}{3} * \pi * r^3$.)

Explain your answer. ______________________

____

____

Date Time

LESSON 9•11

Angle, Perimeter, and Area Problems

Solve each problem. Use substitution to check your answers.

1. $\angle ABC$ is a right angle. What is the degree measure of $\angle CBD$? Of $\angle ABD$?

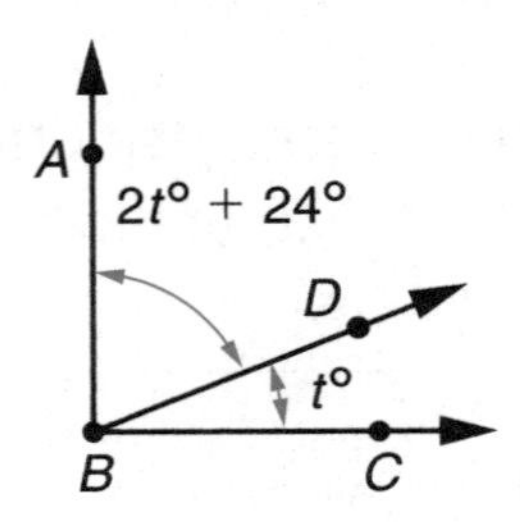

Equation ______________________

Solve.

m$\angle CBD$ = ________ °   m$\angle ABD$ = ________ °

2. Triangle *MJQ* and square *EFGH* have the same perimeter. The dimensions are given in millimeters. What are the lengths of sides *MQ* and *MJ* in triangle *MJQ*?

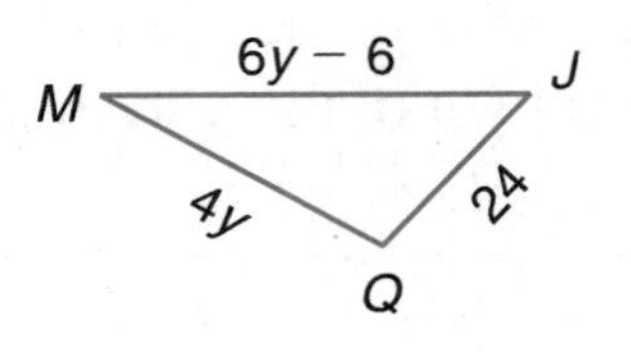

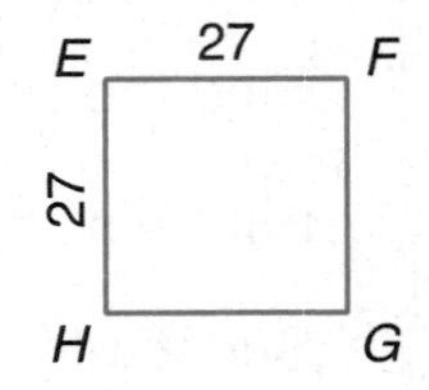

Equation ______________________

Solve.

Length of $\overline{MQ}$ = ________ (unit)   Length of $\overline{MJ}$ = ________ (unit)

3. The area of the shaded part of rectangle *RSTU* is 78 ft^2. Find the length of side *TU*.

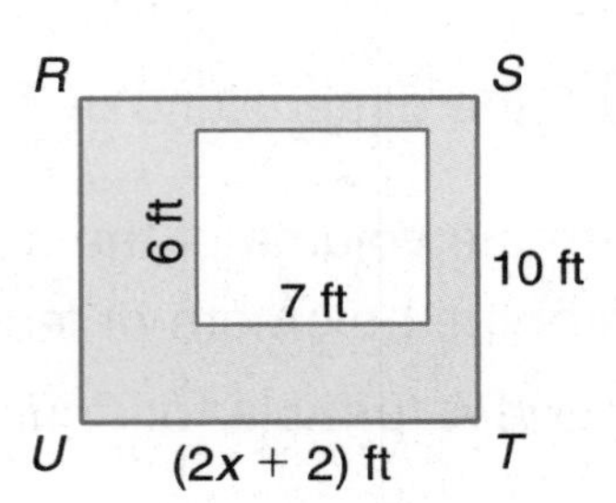

Equation ______________________

Solve.

Length of side *TU* = ________ (unit)

Date   Time

LESSON 9•11

Surface Area of Rectangular Prisms

The surface area (*SA*) of a rectangular prism is the sum of the areas of its faces.

Study the examples below.

Example 1:

Use a net to find the surface area (*SA*) of the rectangular prism.

- Draw and label a net.
- Find the area of each rectangle in the net.

$w = 4$ cm

3 cm ∗ 4 cm = 12 cm²

$l = 6$ cm

6 cm ∗ 3 cm = 18 cm² | 6 cm ∗ 4 cm = 24 cm² | 6 cm ∗ 3 cm = 18 cm² | 6 cm ∗ 4 cm = 24 cm²

$h = 3$ cm

3 cm ∗ 4 cm = 12 cm²

- Find the sum of the areas.

18 + 24 + 18 + 24 + 12 + 12 = 108

Example 2:

Use a formula to find the surface area of the rectangular prism.

- Identify the length (*l*), width (*w*), and height (*h*) of the prism.
- Because there are 3 pairs of opposite faces and opposite faces have the same area, you can use the following formula.

$$\begin{aligned} SA &= 2lw + 2lh + 2wh \\ &= 2(24) + 2(18) + 2(12) \\ &= 48 + 36 + 24 \\ &= 108 \end{aligned}$$

The surface area of the rectangular prism is 108 cm².

Use the formula $SA = 2lw + 2lh + 2wh$ to find the surface area of each rectangular prism.

1.

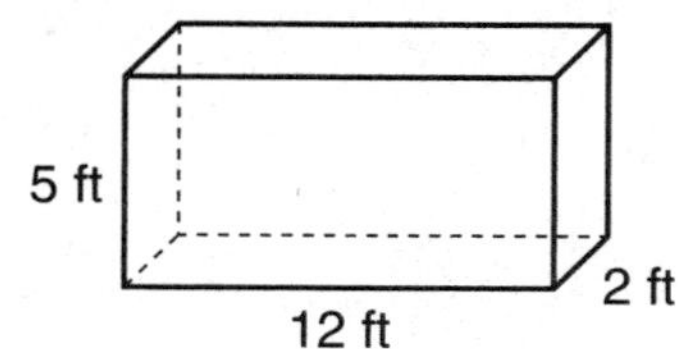

$SA =$ __________ ft²

2.

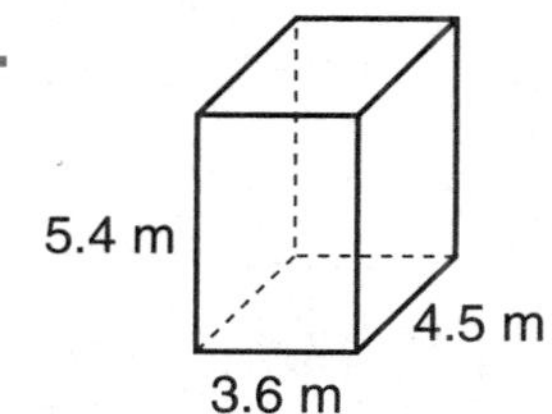

$SA =$ __________ m²

3. Which has the greater surface area, a cube with $s = 8$ m or a rectangular prism with dimensions $l = 6$ m, $w = 5$ m, and $h = 12$ m?

____

____

LESSON 9·11

Surface Area of Cylinders

The surface area (*SA*) of a cylinder is the sum of the areas of its 2 circular bases and the area of its curved surface.

Example:

Find the surface area (*SA*) of the cylinder.

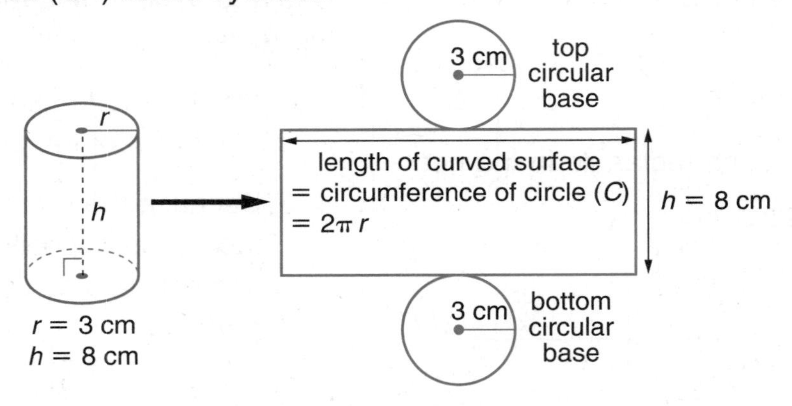

$$
\begin{aligned}
SA &= (2 * \text{area of circular base}) + (\text{area of curved surface}) \\
&= (2 * \text{area of circular base}) + (\text{circumference of base} * \text{cylinder height}) \\
&= (2 * \pi * r^2) + ((2 * \pi * r) * h) \\
&\approx (2 * 3.14 * (3\text{ cm})^2) + ((2 * 3.14 * 3\text{ cm}) * 8\text{ cm}) \\
&\approx (56.52\text{ cm}^2) + (150.72\text{ cm}^2) \\
&\approx 207.24\text{ cm}^2
\end{aligned}
$$

The surface area of the cylinder is 207.24 cm^2.

Use the formula $SA = (2 * \pi * r^2) + ((2 * \pi * r) * h)$ and 3.14 for π to find the surface area of each cylinder.

1.

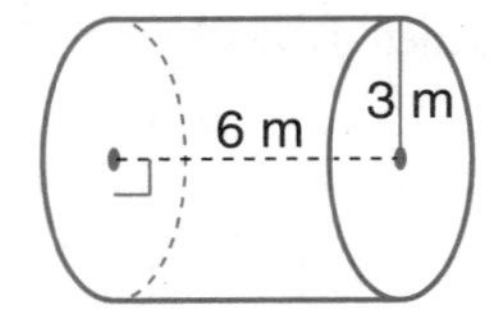

SA = ______________ m^2

2.

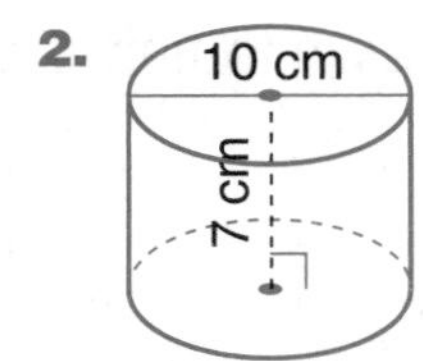

SA = ______________ cm^2

3. A can of tomato juice is a cylinder with a radius of about 7.5 cm and a height of about 20 cm. What is the area of a label that fits around the curved surface of the can with no overlap? Use 3.14 for π.

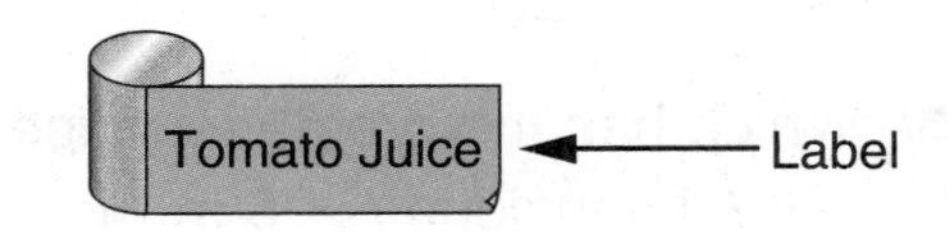

Area of label = ______________ cm^2

Date Time

LESSON 9•11 Math Boxes

1. Find the perimeter of the figure shown below. Use $\frac{22}{7}$ for π.

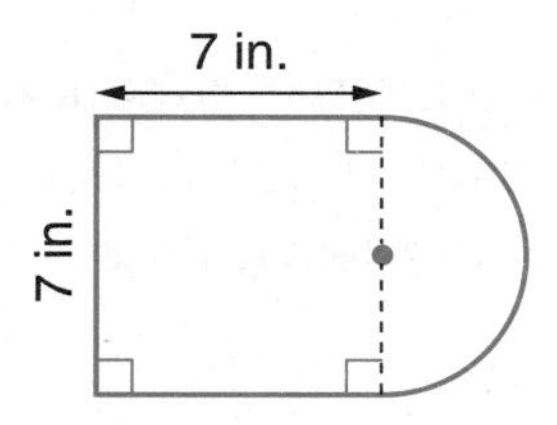

Perimeter = ______ (unit)

2. An aquarium with a rectangular base measuring 36 in. by 12 in. has a volume of 6,912 in.^{3} Find the height (h) of this aquarium.

h = ______ in.

3. Write an equation for each statement. Then solve.

   a. 20% of x is 24.

   Equation ______

   Solution ______

   b. 60% of n is 75.

   Equation ______

   Solution ______

4. Use the order of operations to evaluate each expression.

   a. $15 + 2^2 - 8 \div 4$ ______

   b. $9 * (6 + 2) - (-5)$ ______

   c. $52 - 8 \div 2$ ______

   d. $3[16 - (3 + 7) \div 5]$ ______

5. Fill in each shape to make a recognizable figure. See the example at the right.

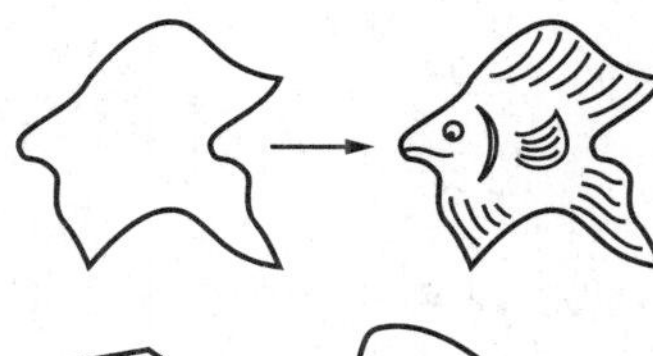

a. 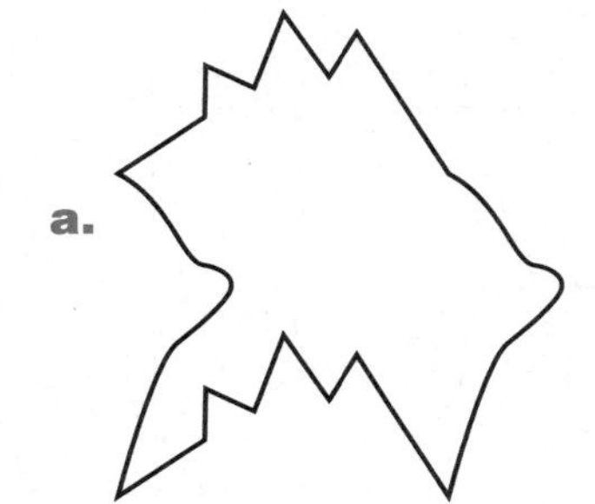

b. 

LESSON 9•12

Squares and Square Roots of Numbers

Math Message

You know that the **square of a number** is equal to the number multiplied by itself. For example, $5^2 = 5 * 5 = 25$.

A **square root** of a number *n* is a number whose square is *n*. For example, a square root of 25 is 5, because $5^2 = 5 * 5 = 25$. A square root of 25 is also -5, because $(-5) * (-5) = (-5)^2 = 25$. Every positive number has 2 square roots, which are opposites of each other.

We use the symbol $\sqrt{\ }$ to write positive square roots. $\sqrt{25}$ is read as *the positive square root of 25.*

1. Write the positive square root of each number.

a. $\sqrt{81}$ = ________ **b.** $\sqrt{100}$ = ________ **c.** $\sqrt{100^2}$ = ________

2. What is the square root of zero? ________

3. Explain why the square root of a negative number does not exist in the real number system.

____

____

To find the positive square root of a number with a calculator, use the [√] key. For example, to find the square root of 25, enter [√] 25 [)] [Enter]. The display will show 5.

4. Use a calculator. Round your answers for **a** and **d** to the nearest hundredth.

a. $\sqrt{17}$ = ________ **b.** $\sqrt{17} * \sqrt{17}$ = ________ **c.** $\sqrt{\pi} * \sqrt{\pi}$ = ________

d. $\sqrt{\pi}$ = ________ **e.** $(\sqrt{17})^2$ = ________ **f.** $\sqrt{\frac{1}{16}}$ = ________

5. The length of a side of a square is $\sqrt{6.25}$ centimeters.
What is the area of the square? ________ (unit)

6. The area of a square is 21 square inches.
What is the length of a side to the nearest tenth of an inch? ________ (unit)

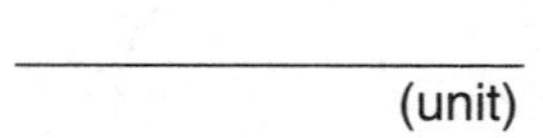

7. The radius of a circle is $\sqrt{20}$ feet.
What is its area to the nearest tenth of a foot? About ________ (unit)

Date Time

LESSON 9•12 Verifying the Pythagorean Theorem

In a right triangle, the side opposite the right angle is called the **hypotenuse.** The other two sides are called the **legs of the triangle.**

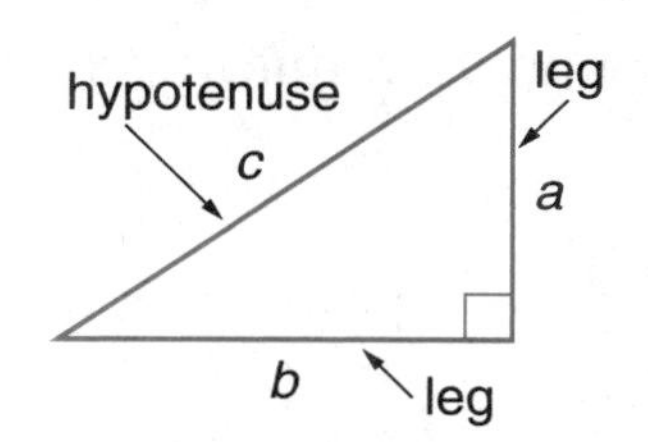

Think about the following statement:
If *a* and *b* are the lengths of the legs of a right triangle and *c* is the length of the hypotenuse, then $a^2 + b^2 = c^2$.

This statement is known as the **Pythagorean theorem.**

1. To verify that the Pythagorean theorem is true, use a blank sheet of paper that has square corners. Draw diagonal lines to form 4 right triangles, one at each corner. Then measure the lengths of the legs and the hypotenuse of each right triangle, to the nearest millimeter. Record the lengths in the table below. Then complete the table.

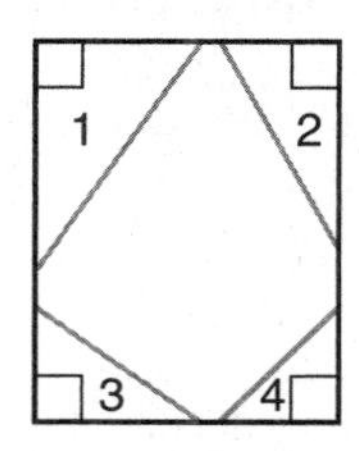

Triangle	Leg (a)	Leg (b)	Hypotenuse (c)	$a^2 + b^2$	c^2
1					
2					
3					
4					

2. Compare $(a^2 + b^2)$ to c^2 for each of the triangles you drew. Why might these two numbers be slightly different?

____

____

____

3. Use the Pythagorean theorem to find c^2 for the triangle at the right. Then find the length *c*.

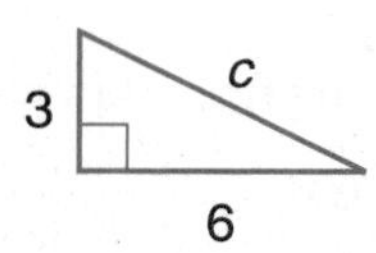

$c^2 =$ __________ units^2 *c* is about __________ units.

Date ____________________ Time ____________________

LESSON 9·12 Using the Pythagorean Theorem

In Problems 1–6, use the Pythagorean theorem ($a^2 + b^2 = c^2$) to find each missing length. Round your answer to the nearest tenth.

1.

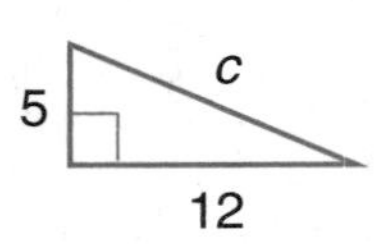

Equation $c^2 = 5^2 + 12^2$

$c^2 =$ ________ $c =$ ________

2.

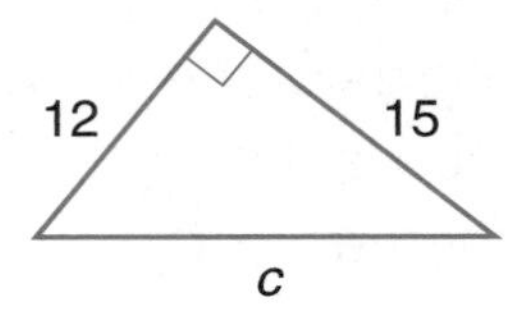

Equation ____________________

$c^2 =$ ________ $c =$ ________

3.

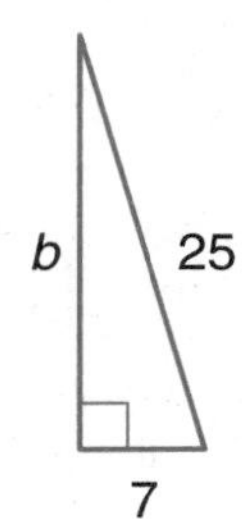

Equation ____________________

$b^2 =$ ________ $b =$ ________

4.

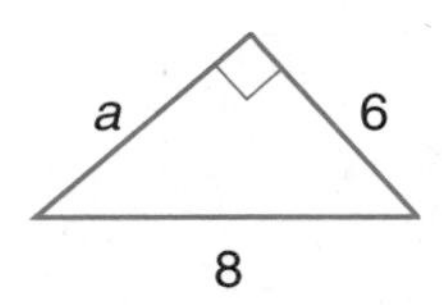

Equation ____________________

$a^2 =$ ________ $a =$ ________

5.

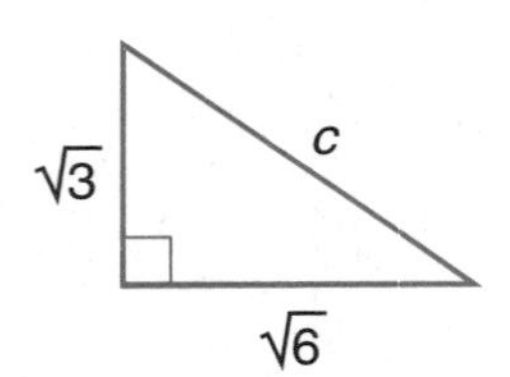

Equation ____________________

$c^2 =$ ________ $c =$ ________

6.

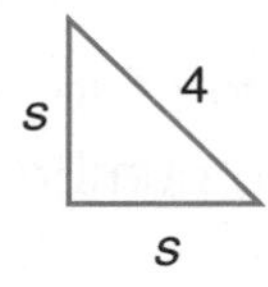

Equation ____________________

$s^2 =$ ________ $s =$ ________

7. Is the triangle shown a right triangle? ________

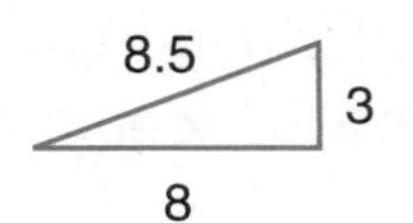

Explain. ____________________

____

____

____

Date Time

LESSON 9•12

Math Boxes

1. What is the area of the shaded region in the figure shown below?

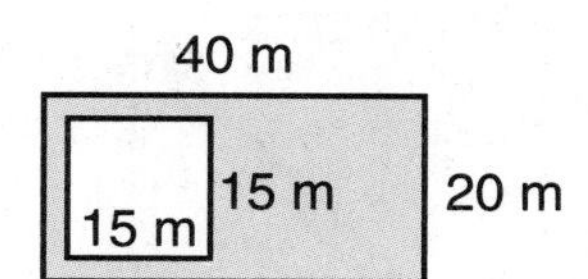

Area = ______ (unit)

2. Solve.

a. $w * 10^{-2} = 28.2$ ______

b. $420 * k = 140$ ______

c. $\frac{5}{2} - p = \frac{7}{4}$ ______

d. $(\sqrt{11})^2 = n$ ______

3. The table at the right shows the approximate number of calories a 150-pound person uses per hour while performing various activities.

For which activity does the person use about $\frac{2}{3}$ the number of calories used in jumping rope? Fill in the circle next to the best answer.

○ **A.** swimming

○ **B.** walking

○ **C.** volleyball

○ **D.** basketball

Activity	Calories Per Hour
Swimming (25 yd/min)	275
Walking (3 mph)	320
Volleyball	350
Basketball	500
Jumping rope	750
Running (7 mph)	920

4. Measure the angles.

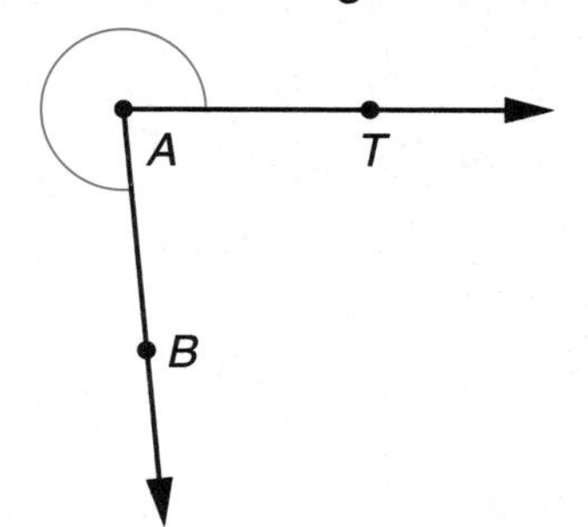

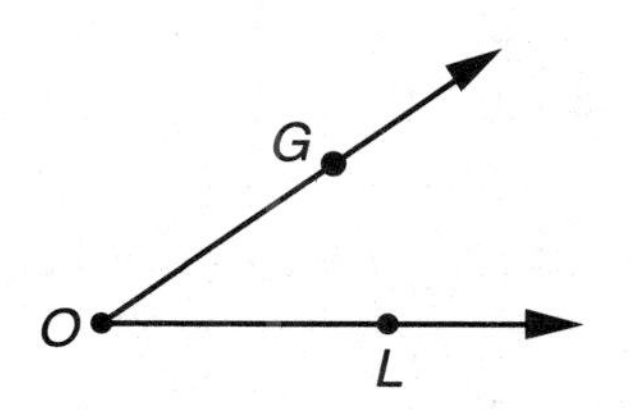

a. Reflex $\angle BAT$ measures ______°.

b. $m\angle LOG$ is about ______°.

LESSON 9•13

Similar Figures and the Size-Change Factor

The 2 butterfly clamps shown below are similar because they each have the same shape. One clamp is an enlargement of the other. The size-change factor tells the amount of enlargement.

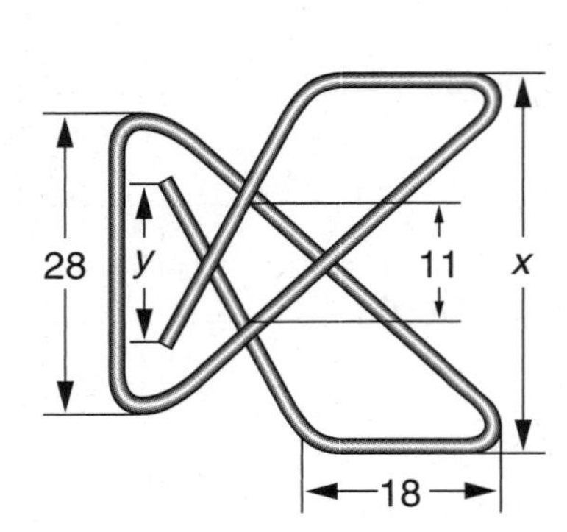

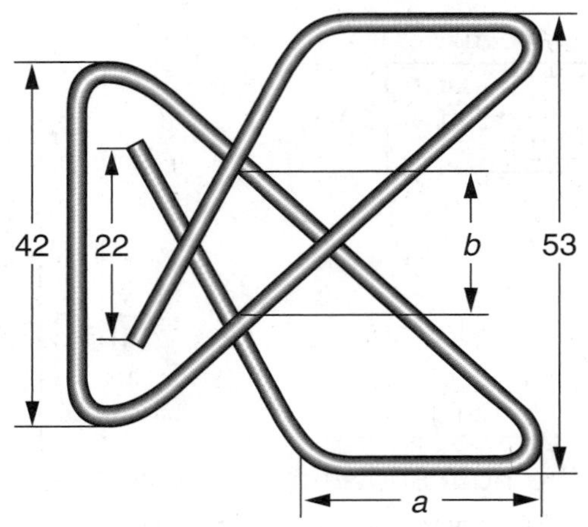

unit: millimeters (mm)

1. The size-change factor for the clamps shown above is ______________.

In Problems 2–5, use the size-change factor to find the missing lengths.

2. a = ______________ mm = ______________ cm

3. b = ______________ mm = ______________ cm

4. x = ______________ mm = ______________ cm

5. y = ______________ mm = ______________ cm

6. If a butterfly clamp is straightened, it forms a long, thin cylinder. When the small clamp is straightened, it is 21 cm long, and the thickness (diameter) of the clamp is 0.15 cm. Its radius is 0.075 cm. Calculate the volume of the small clamp. Use the formula $V = \pi r^2 h$.

Volume of small clamp = ______________ cm^3 (to the nearest thousandth cm^3)

7. Find the length, thickness (diameter), and volume of the large clamp.

Length = ______________ cm Diameter = ______________ cm

Volume of large clamp = ______________ cm^3 (to the nearest thousandth cm^3)

Date ______ Time ______

LESSON 9•13

Indirect Measurement Problems

In the following problems, you will use indirect measurement to determine the heights and lengths of objects that you cannot directly measure.

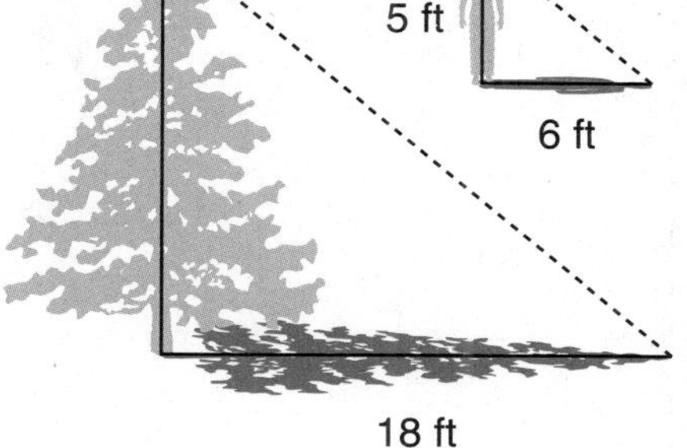

1. A tree is too tall to measure, but it casts a shadow that is 18 feet long. Ike is standing near the tree. He is 5 feet tall and casts a shadow that is 6 feet long.

The light rays, the tree, and its shadow form a triangle that is **similar** to the triangle formed by the light rays, Ike, and his shadow.

What is the size-change factor of the triangles? ____________

About how tall is the tree? ____________

2. Ike's dad is 6 feet tall. He is standing near the Washington Monument, which is 555 feet tall. Ike's dad casts a 7-foot shadow. About how long a shadow does the Washington Monument cast? (*Hint:* Draw sketches that include the above information.) ____________

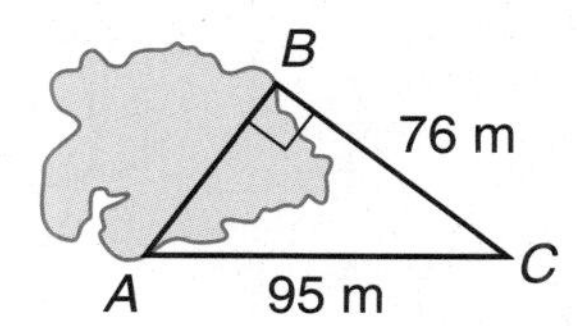

3. A surveyor wants to find the distance between points *A* and *B* on opposite ends of a lake. He sets a stake at point *C* so that angle *ABC* is a right angle. By measuring, he finds that $\overline{AC}$ is 95 meters long and $\overline{BC}$ is 76 meters long.

How far across the lake is it from point *A* to point *B*? ____________

Date Time

LESSON 9•13 Math Boxes

1. Find the area of the shaded region in the figure below. Use $\frac{22}{7}$ for π.

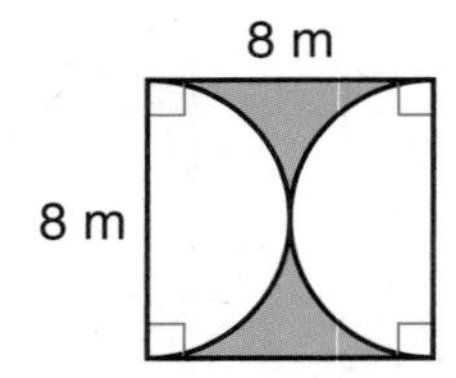

Area = ______ (unit)

2. Which expression represents the volume of the cylinder shown below?
Circle the best answer.

A 800π

B 200π

C 80π

D 40π

3. Write an equation for each statement. Then solve.

a. 125% of x is 625.

Equation ______

Solution ______

b. n% of 76 is 19.

Equation ______

Solution ______

4. Use the order of operations to evaluate each expression.

a. $(-3)^2 + 5 * 2^3$ ______

b. $(-1)^{50} * (-1)^{49}$ ______

c. $-6 + 2 * 3^3$ ______

d. $(-1)^5 * (2^4 - 13)^2$ ______

5. Fill in each shape to make a recognizable figure. See the example at the right.

a.

b.

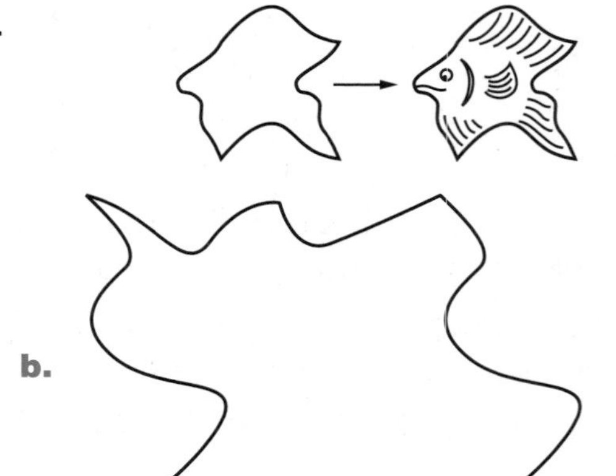

Date Time

LESSON 9•14 Math Boxes

1. Draw and label an obtuse angle *HIJ*. Then measure it.

m∠*HIJ* is about ________°.

2. Which of the hexagons shown below is a regular hexagon? Circle one.

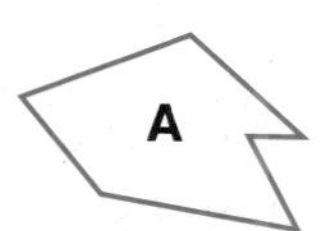

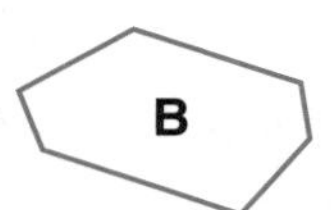

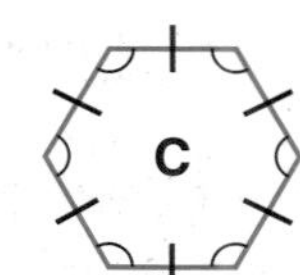

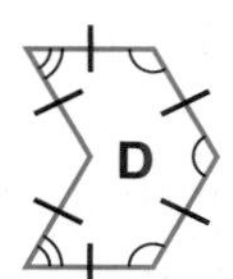

3. Use the diagonals to find the sum of the interior angle measures of the figure shown below.

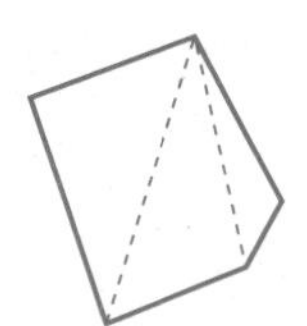

Sum of the interior angle measures = ________°.

4. Draw lines of symmetry for each figure shown below.

a.

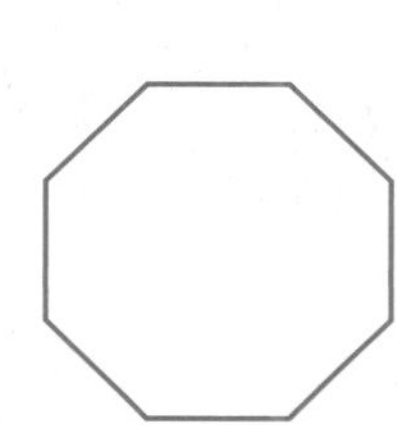

b.

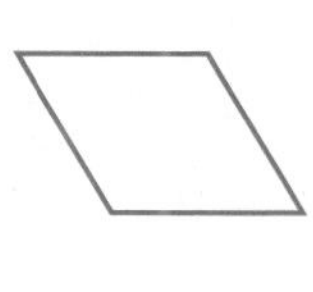

5. Identify whether the preimage (1) and image (2) are related by a translation, a reflection, or a rotation.

Write your answer on the line below.

____

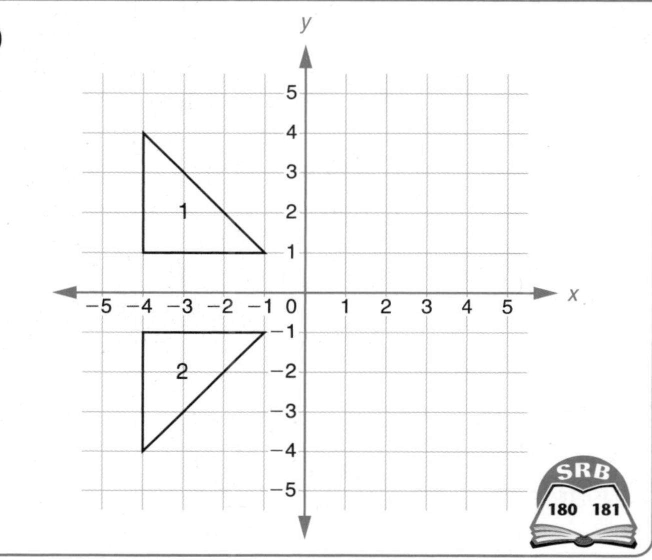

Date Time

Semiregular Tessellations

A **semiregular tessellation** is made up of two or more kinds of regular polygons. In a semiregular tessellation, the arrangement of angles around each **vertex point** looks the same. There are exactly 8 different semiregular tessellations. One is shown below.

Find and draw the other 7 semiregular tessellations. The only polygons that are possible in semiregular tessellations are equilateral triangles, squares, regular hexagons, regular octagons, and regular dodecagons. Use your Geometry Template and the template of a regular dodecagon that your teacher will provide.

Experiment first on a separate sheet of paper. Then draw the tessellations below and on the next page. Write the name of each tessellation.

1.

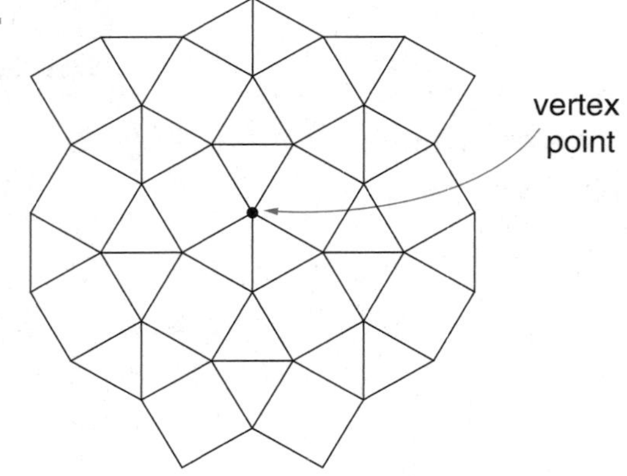

Name 3.3.4.3.4

2.

Name ______________

3.

Name ______________

4.

Name ______________

Date Time

Semiregular Tessellations *continued*

5.

Name ____________________

6.

Name ____________________

7.

Name ____________________

8.

Name ____________________

Date Time

LESSON 10•1

Math Boxes

1. You can use the formula $s = 180 * (n - 2)$ to find the sum of the interior angle measures of a polygon having n sides.

   For example, the sum of the interior angle measures (s) of a 7-sided polygon (n) is $180 * (7 - 2) = 180 * 5 = 900$, or 900°.

   Suppose the sum of the interior angle measures of a polygon is 1,800°. How many sides does this polygon have?

    sides

2. Without using a protractor, find the measure of each numbered angle.

E F D G C H P 1 2 108° 38°

m∠1 = ______ m∠2 = ______

3. Multiply or divide. Write your answers in simplest form.

   a. $3\frac{8}{9} * 4\frac{5}{6} =$ ______

   b. ______ $= \frac{1}{5} * \frac{38}{3}$

   c. ______ $= \frac{24}{15} \div \frac{1}{2}$

   d. ______ $= \frac{3}{7} * \frac{22}{3}$

   e. $\frac{24}{8} \div \frac{12}{7} =$ ______

4. Complete the table. Write a number sentence to describe the relationship between the numbers in the table.

x	*y*
12	−4
6	
	$-\frac{1}{3}$
−3	1
	6

Number sentence ______

5. Write an expression for the perimeter of the figure. Combine like terms.

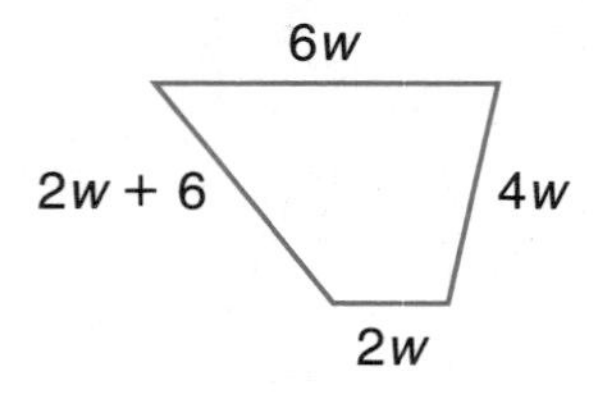

Perimeter = ______

6. Tell whether each of the following is true or false.

   a. 1 cm > 1 in. ______

   b. 1 m > 1 yd ______

   c. 200 cm > 1 ft ______

   d. 10 mm > 1 in. ______

   e. 5 ft < 100 cm ______

Date Time

LESSON 10•2

My Tessellation

On a separate sheet of paper, create an Escher-type translation tessellation using the procedure described on page 360 of the *Student Reference Book.* Experiment with several tessellations until you create one that you especially like.

Trace your final tessellation template in the space below.

In the space below, use your tessellation template to record what your tessellation looks like. Add details or color to your final design.

LESSON 10•2 Translations, Reflections, Rotations

When a figure is translated, reflected, or rotated, its size and shape remain the same; only its position changes.

Consider the shaded right triangle whose vertices are $(-1,1)$, $(-5,1)$, and $(-1,4)$.

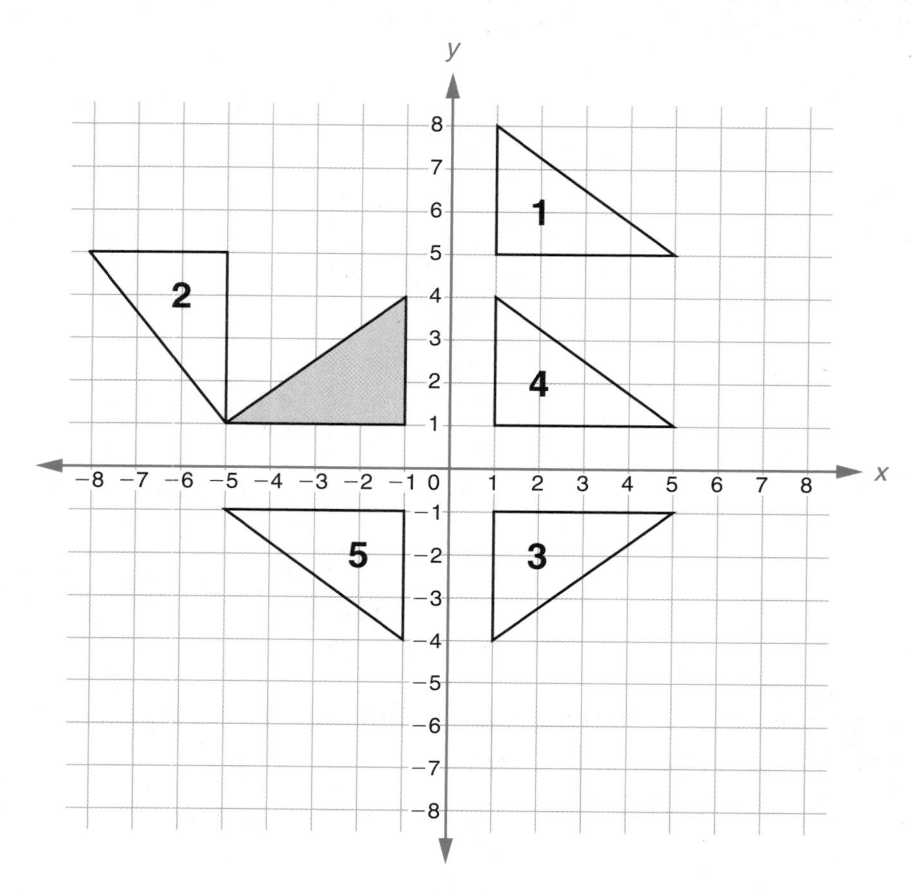

Identify which triangle results from performing the following transformations on the shaded right triangle.

a. ______ A reflection over the *y*-axis

b. ______ A 90° counterclockwise rotation around point $(-5,1)$

c. ______ A reflection over the *x*-axis

d. ______ A reflection over the *y*-axis, followed by a reflection over the *x*-axis

e. ______ A reflection over the *y*-axis, followed by a translation 4 units up

Date ______________________ Time ______________________

LESSON 10•2

Math Boxes

1. Apply the order of operations to evaluate each expression.

a. $15 - 3.3 * 4 =$ ____________

b. $\frac{20}{4} * 5 + (-8) * 2 =$ ____________

c. $0.01 + 0.01 * 10 + 0.01 =$ ____________

d. $8 * (2 + -5) - 4 =$ ____________

e. $7 * 3^2 - \frac{10}{2} =$ ____________

2. Estimate each quotient or product.

a. $5.25 \div 2.003$ About ____________

b. $4.29 * 67.1$ About ____________

c. $80.25 \div 18.93$ About ____________

d. $52.31 * 19.9$ About ____________

3. Convert between number-and-word and standard notations.

a. 2,500,000 ______________________

b. 0.3 thousand ______________________

c. 7,400,000,000,000

____

d. 1,234.5 million

____

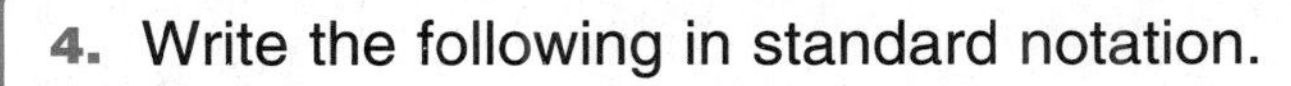

4. Write the following in standard notation.

a. $5.38 * 10^7 =$ ______________________

b. $6.91 * 10^{-3} =$ ______________________

c. $3.04 * 10^0 =$ ______________________

d. ______________________ $= 9.9011 * 10^5$

e. $7.2 * 10^{-6} =$ ______________________

5. Write an algebraic expression for the following situation.

Rafael is twice as old as his brother Jorge was 3 years ago. Jorge is *j* years old now. How old is Rafael?

Expression ______________________

6. Fill in the blanks to complete each number sentence.

a. 8(40 + 5) = (____ ∗ 40) + (8 ∗ ____)

b. (10 − 3)6 = (10 ∗ ____) − (3 ∗ ____)

c. (9 ∗ 50) + (9 ∗ ____) = 9(____ + 4)

d. (7 ∗ 20) − (7 ∗ 3) = ____(____ − ____)

Date Time

LESSON 10•3 Rotation Symmetry

Cut out the figures on Activity Sheet 7. Cut along the dashed lines only. Using the procedure demonstrated by your teacher, determine the number of different ways in which you can rotate (but not flip) each figure so the image exactly matches the preimage. Record the order of rotation symmetry for each figure.

1.

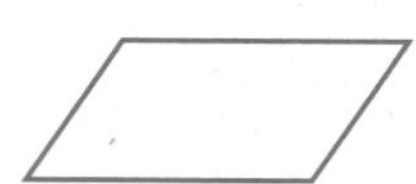

Order of rotation symmetry ________

2.

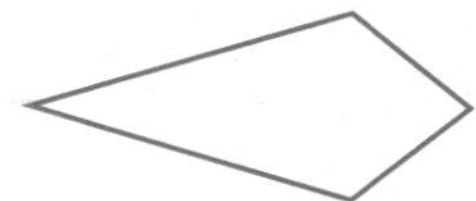

Order of rotation symmetry ________

3.

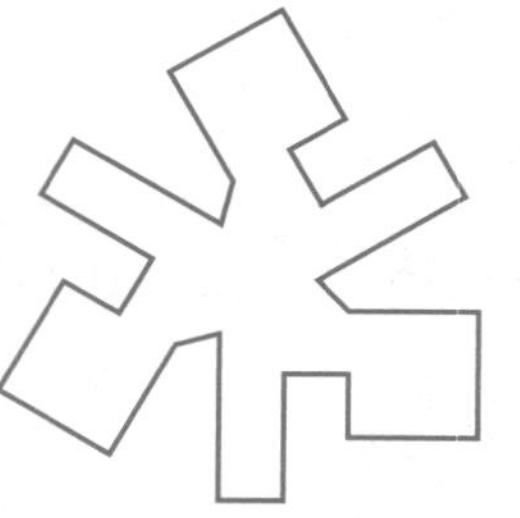

Order of rotation symmetry ________

4.

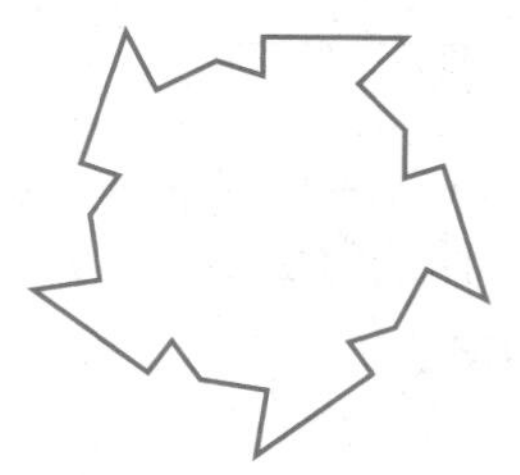

Order of rotation symmetry ________

Try This

5. The 10 of hearts has point symmetry. When the card is rotated 180°, it looks the same as it did in the original position.

original position

180° rotation

The 9 of spades does not have point symmetry. When the card is rotated 180°, it does not look the same as it did in the original position.

original position

180° rotation

Which of the cards in an ordinary deck of playing cards (not including face cards) have point symmetry? ________________________________

____

____

To learn a magic trick that uses point symmetry with playing cards, see page 355 in the *Student Reference Book.*

Date Time

LESSON 10•3 Math Boxes

1. You can use the formula $s = 180 * (n - 2)$ to find the sum of the interior angle measures of a polygon having n sides.

   Find the value of x in the figure below.

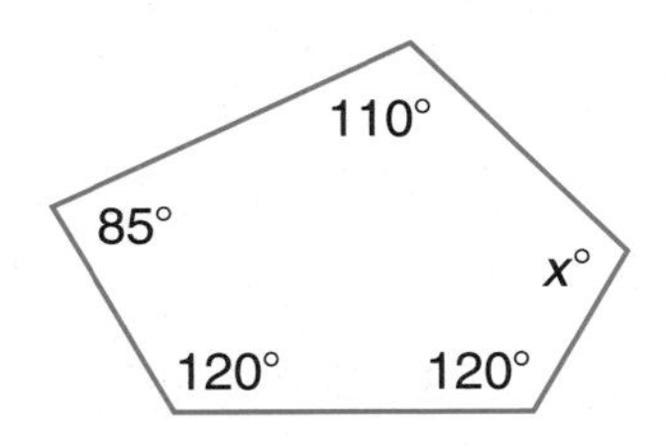

$x =$ ______ °

2. Write an equation that describes the angle relationships shown. Then solve it.

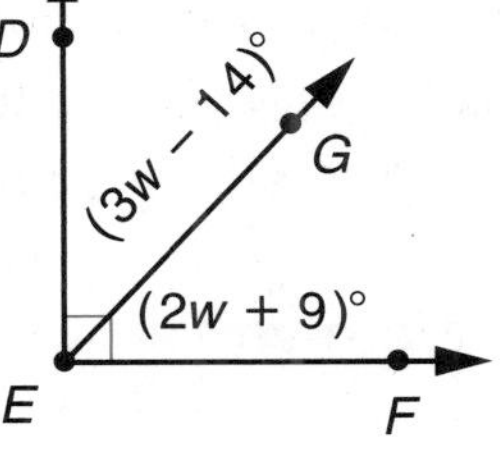

Equation ______

$w =$ ______

m∠$DEG =$ ______ °

m∠$GEF =$ ______ °

3. Multiply or divide. Write your answers in simplest form.

   a. $1\frac{3}{7} * 2\frac{1}{5} =$ ______

   b. ______ $= 3\frac{6}{8} * \frac{28}{6}$

   c. ______ $= 5\frac{1}{10} \div 2\frac{5}{4}$

   d. ______ $= \frac{46}{3} \div 20$

   e. $5\frac{3}{8} * \frac{1}{8} =$ ______

4. Complete the table. Write a number sentence to describe the relationship between the numbers in the table.

x	y
100	
30	$\frac{3}{2}$
10	
−2	$-\frac{1}{10}$
	$-\frac{3}{4}$

Number sentence ______

5. The area of the parallelogram shown below is 63 cm^2. Which equation can you use to find its height? Circle the best answer.

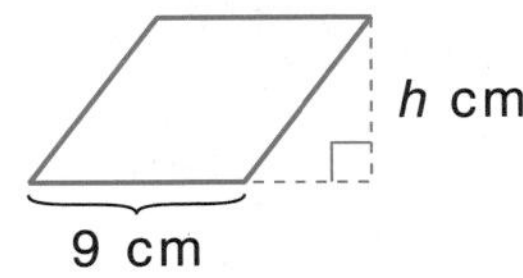

A. $\frac{9h}{2} = 63$ **B.** $2h + 18 = 63$

C. $9h = 63$ **D.** $\frac{9}{63} = h$

6. Tell whether each of the following is true or false.

   a. 1 L > 1 pt ______

   b. 1 mL < 1 fl oz ______

   c. 1 L > 1 gal ______

   d. 1 kg > 1 lb ______

   e. 1 g < 1 oz ______

Date Time

LESSON 10•4 Shrinking Quarter

Math Message

1. Carefully trace the dime-size circle below onto the center of an $8\frac{1}{2}$ in. by 11 in. sheet of paper. Cut out the circle.
   Try to slip a quarter through the hole in the page without tearing the paper.

2. Were you able to do this trick? If you were, explain how you did it. If not, explain why you could not.

____

____

____

____

____

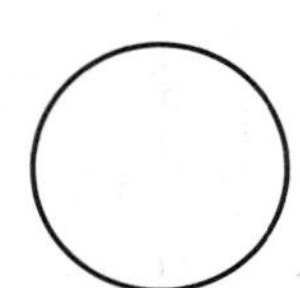

Date Time

LESSON 10•4

Math Boxes

1. Apply the order of operations to evaluate each expression.

a. $4 * \frac{7}{2} + 7 =$ ______

b. $8 + (-15) * 6 =$ ______

c. $\frac{6^2}{9} + 3 * 4 =$ ______

d. $8 + 7 - (-2) * 5 =$ ______

e. $12 / 6 + 9 * 3 =$ ______

2. Estimate each quotient or product.

a. $44.2 * 37$ About ______

b. $708 \div 0.52$ About ______

c. $625.7 \div 8.3$ About ______

d. $99.4 * 3.7$ About ______

3. Convert between number-and-word and standard notations.

a. 6,500,000 ______

b. 0.75 billion ______

c. 12,500,000,000,000

______

d. 57.25 million

______

4. Write the following in standard notation.

a. $2.73 * 10^5 =$ ______

b. $1.03 * 10^{-4} =$ ______

c. $9.855 * 10^0 =$ ______

d. ______ $= 4.226 * 10^6$

e. $5.435 * 10^{-2} =$ ______

5. Chen ran 2 miles more than two-thirds as far as Kayla ran. If Kayla ran k miles, how many miles did Chen run?

Which algebraic expression can you use to answer the question? Fill in the circle next to the best answer.

Ⓐ $\frac{2}{3}k + 2$ Ⓑ $k + \frac{2}{3}$

Ⓒ  $2k + \frac{2}{3}$ Ⓓ $2k \div 3$

6. Fill in the blanks to complete each number sentence.

a. $9(20 + 7) = (\_\_\_\_ * 20) + (9 * \_\_\_\_)$

b. $(50 - 5)3 = (50 * \_\_\_\_) - (5 * \_\_\_\_)$

c. $(8 * 70) + (8 * \_\_\_\_) = 8(\_\_\_\_ + 7)$

d. $(6 * 50) - (6 * 9) = \_\_\_\_(\_\_\_\_ - \_\_\_\_)$

Date Time

LESSON 10•4

Rubber-Sheet Geometry

You and your partner will need the following materials: 3 latex gloves, a straightedge, a pair of scissors, and a permanent marker.

Step 1 Cut the fingers and thumb off of each glove.

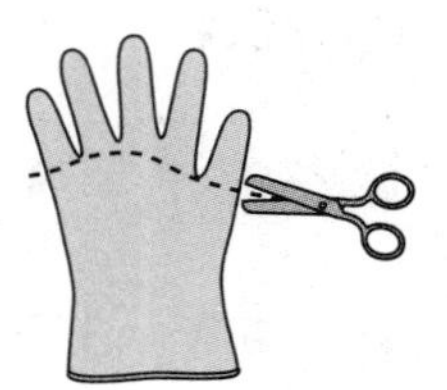

Step 2 Make a vertical cut through each of the three cylinders that remain. The cutting creates 3 rubber sheets.

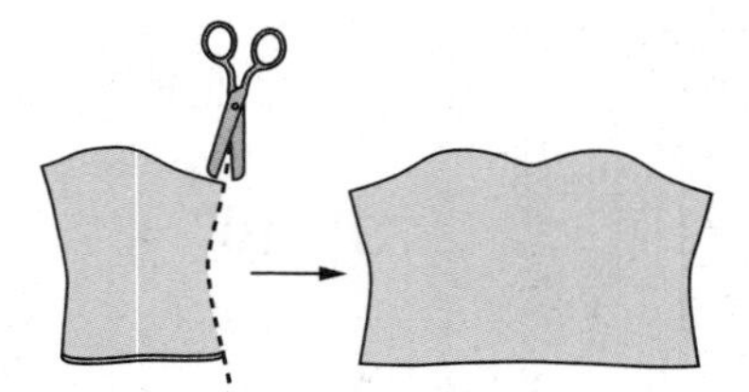

Step 3 Use a permanent marker and a straightedge to draw the following figures on the rubber sheets. Draw the figures large enough to fill most of the sheet.

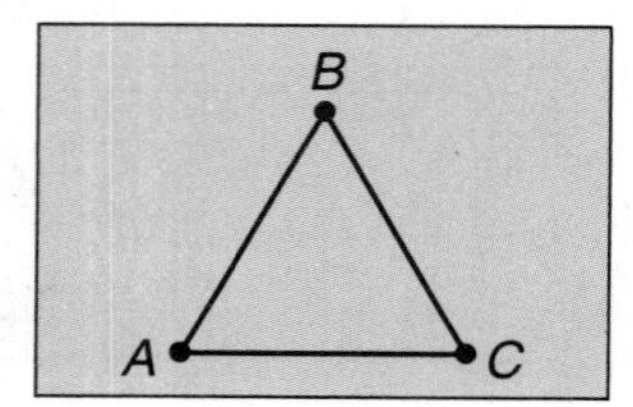

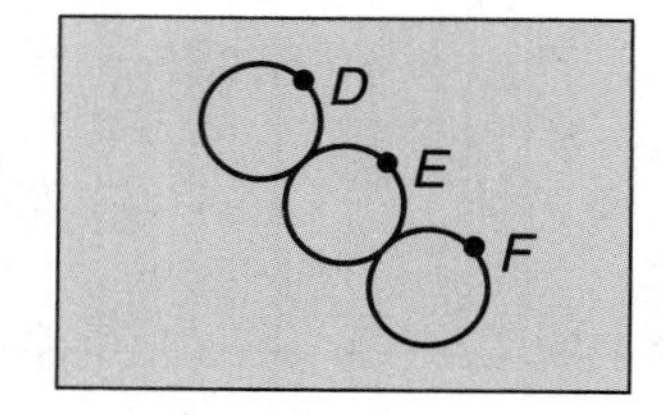

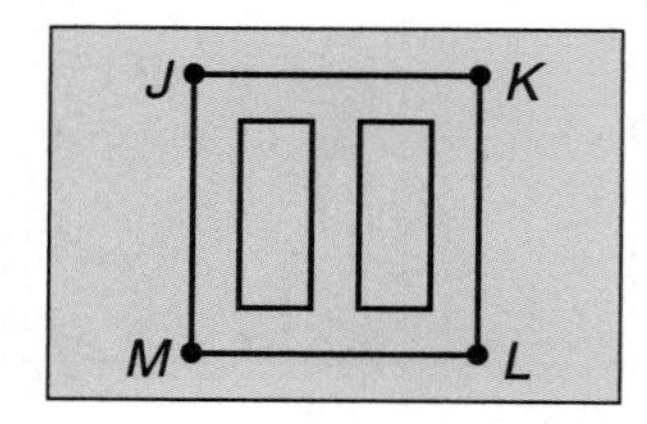

Step 4 Work with your partner to stretch the rubber sheets to see what other figures you can make.

Step 5 Complete journal page 381.

Date Time

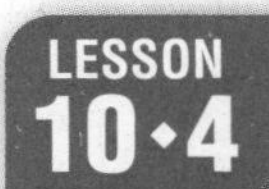

Rubber-Sheet Geometry *continued*

1. Experiment with the figures on your rubber sheets. Circle any of the transformed figures in the right column that are topologically equivalent to the corresponding original figure in the left column.

Original Figure	Transformed Figures

2. Choose one of the above figures that you did not circle and explain why it is not topologically equivalent to the original figure.

____

____

____

Date Time

LESSON 10•5

Math Boxes

1. Draw the line(s) of reflection symmetry for the figure below. Then determine its order of rotation symmetry.

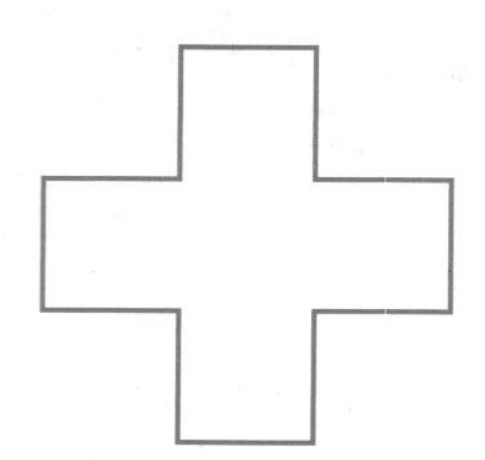

Order of rotation symmetry ______

2. Lines p and q are parallel. Write an equation you can use to find the value of x.

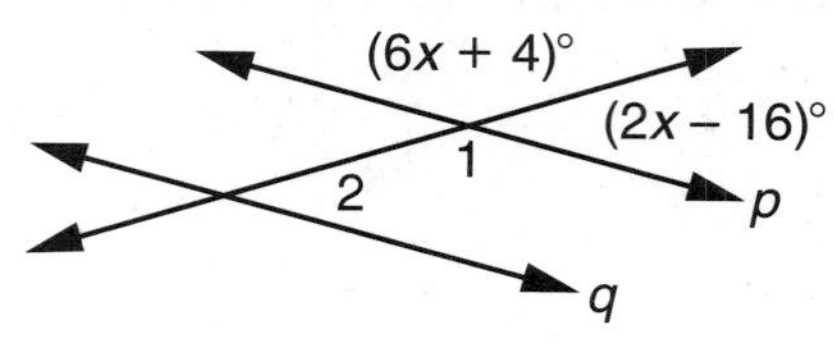

Equation ______

x = ______ °

m∠1 = ______ °

m∠2 = ______ °

3. Multiply or divide. Write your answers in simplest form.

a. $1\frac{3}{5} * (-2\frac{1}{2}) =$ ______

b. ______ $= (-\frac{7}{12})(-\frac{3}{84})$

c. ______ $= 5\frac{1}{10} \div 2\frac{5}{4}$

d. ______ $= -1\frac{1}{3} \div (-\frac{5}{9})$

e. $-3\frac{2}{3} \div (-2\frac{4}{9}) =$ ______

4. Complete the table. Write a number sentence to describe the relationship between the numbers in the table.

x	y
10	23
4	
$\frac{1}{2}$	4
	3
−3	−3

Number sentence ______

5. Tennis balls with a diameter of 2.5 in. are packaged 3 to a can. The can is a cylinder. Find the volume of the space in the can that is *not* occupied by tennis balls. Assume that the balls touch the sides, top, and bottom of the can.

Use the formula $V = \pi r^3$ and 3.14 for π. Round your answer to the nearest hundredth.

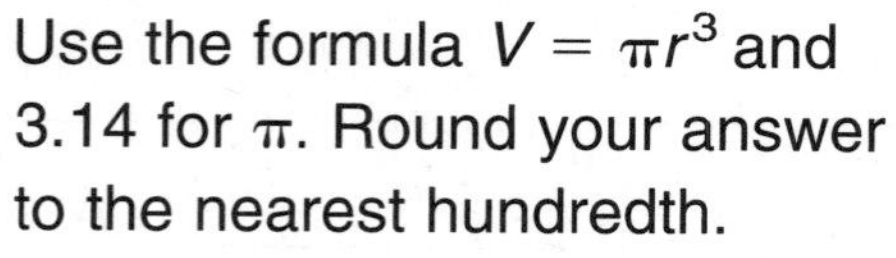

Volume (not occupied by balls) = ______ (unit)

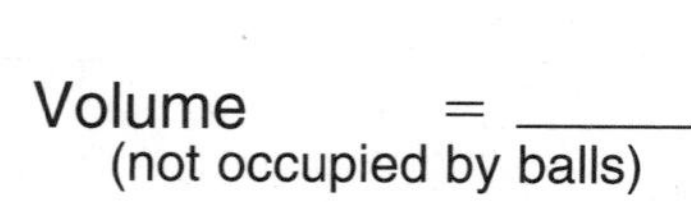

6. Complete.

a. 1 ft^2 = ______ $in.^2$

b. 1 m^2 = ______ cm^2

c. 1 yd^2 = ______ ft^2

d. 1 ft^3 = ______ $in.^3$

e. 1 yd^3 = ______ ft^3

Date Time

Making a Möbius Strip

Follow the steps below to make a Möbius strip.

Materials

- ☐ a sheet of newspaper or adding machine tape
- ☐ scissors
- ☐ tape
- ☐ a bright color crayon, marker, or pencil

Step 1 Cut a strip of newspaper about $1\frac{1}{2}$ inches wide and as long as possible, or cut a strip of adding machine tape about 2 feet long.

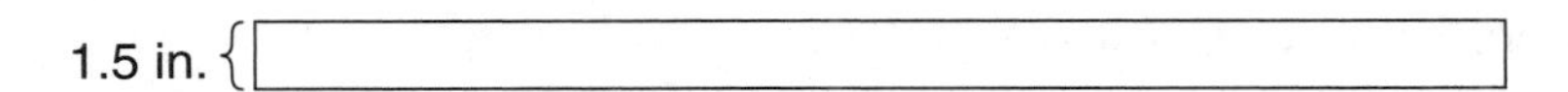

Step 2 Put the ends of the strip together as though you were making a simple loop.

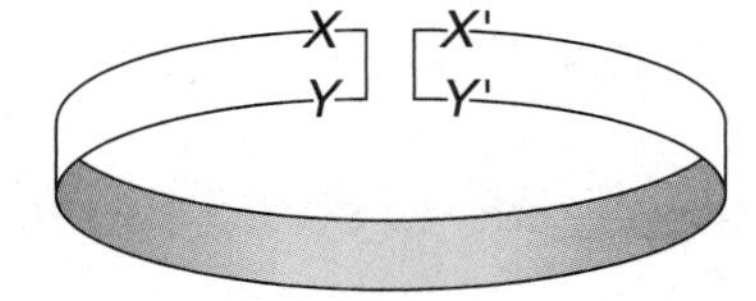

Step 3 Give one end of the strip a half-twist and tape the two ends together.

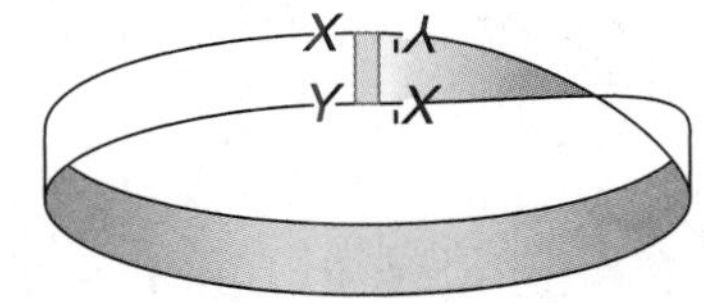

You have just made what mathematicians call a **Möbius strip.** How is it different from the simple loop of paper used in the Math Message? Do you notice anything special about it?

____

____

____

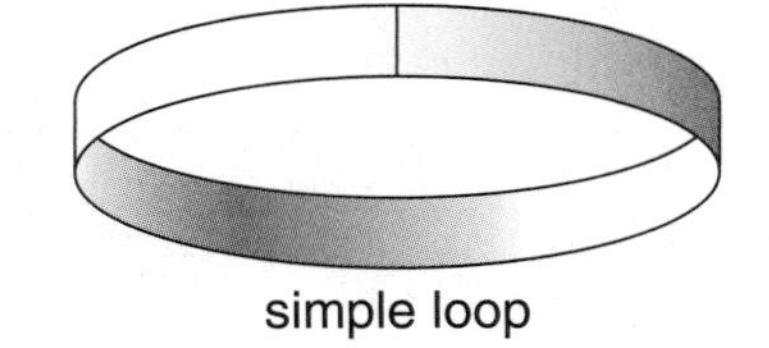
simple loop

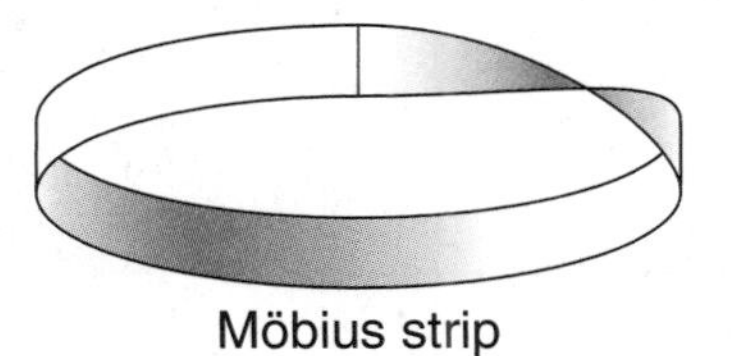
Möbius strip

Date _____________ Time _____________

LESSON 10•5

Experimenting with Möbius Strips

1. How many sides do you think your Möbius strip has? _________ side(s)

2. Use a marker to shade one side of your Möbius strip.

3. Now how many sides do you think your Möbius strip has? Explain.

____

4. How many edges do you think your Möbius strip has? _________ edge(s)

5. Use your marker to color one edge of your Möbius strip.

6. Now how many edges do you think your Möbius strip has? Explain.

____

____

Cutting Möbius strips leads to some surprising results.

7. Predict what will happen if you cut your Möbius strip in half lengthwise.

____

____

8. Now cut your Möbius strip in half lengthwise. How many strips did you get? _________ strip(s)

Compare the lengths and widths of the new strip and the original strip.

Describe your observations. ____

____

How many half-twists does your new strip have? _________ half-twist(s)

9. Make another Möbius strip and cut it one-third of the way from the edge. You may find it helpful to draw lines on the strip before cutting.

What happened? ____

____

____

LESSON 10•5 Experimenting with Möbius Strips *continued*

10. Make another Möbius strip and a simple loop. Then tape the loop and the Möbius strip together at right angles.

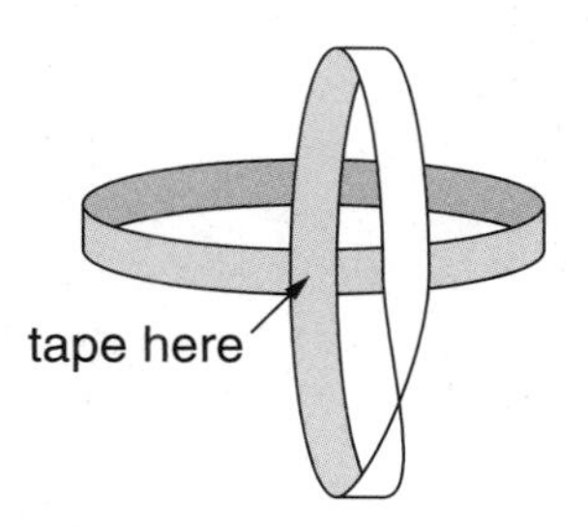

Cut the Möbius strip and the loop in half lengthwise. What happened?

____

____

11. Experiment with cutting Möbius strips in half and in thirds lengthwise. Try putting two or more half-twists in the band before you tape it. Describe what you did, as well as your results.

____

____

____

____

12. Shortly after the first Earth Day on April 22, 1970, a producer of recycled paperboard sponsored a contest. Contestants presented designs that symbolized the company's recycling process. Gary Anderson, a student at the University of Southern California, won the top prize. Anderson's symbol was a three-chasing-arrows Möbius strip.

Explain what you think Anderson was trying to say about recycling with his symbol.

____

____

Date Time

LESSON 10•5

Reflection and Rotation Symmetry

For each figure, draw the line(s) of reflection symmetry, if any. Then determine the order of rotation symmetry for the figure.

1.

Order of rotation symmetry ______

2.

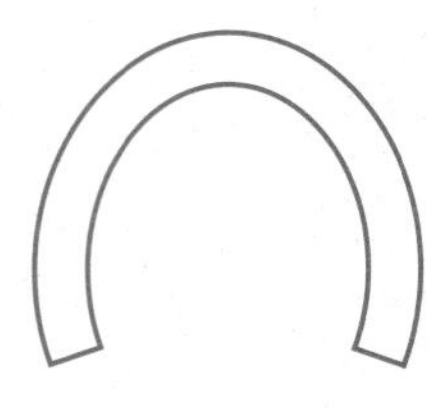

Order of rotation symmetry ______

3.

Order of rotation symmetry ______

4.

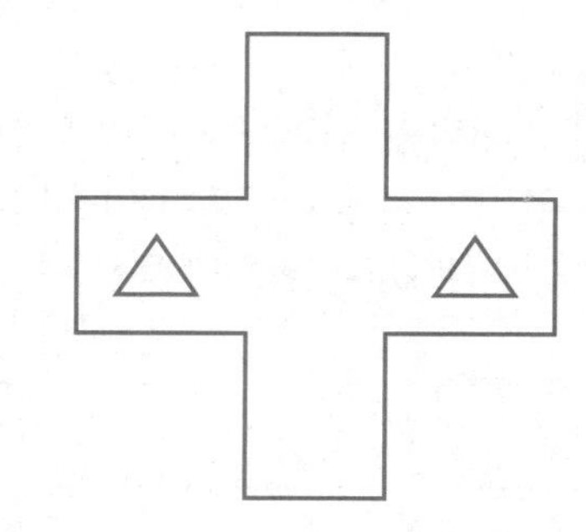

Order of rotation symmetry ______

5.

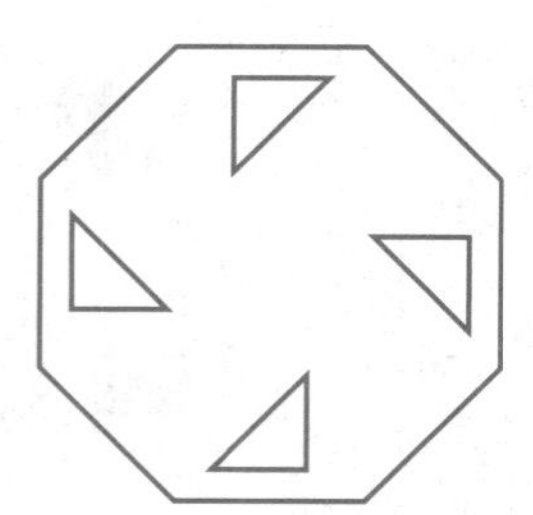

Order of rotation symmetry ______

6.

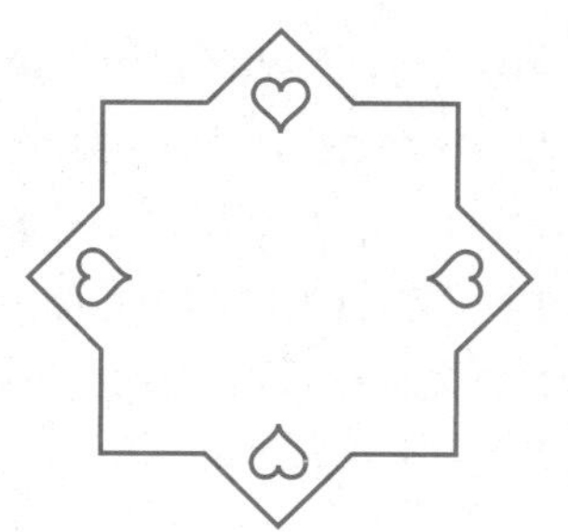

Order of rotation symmetry ______

Date Time

Reference

Metric System

Units of Length

1 kilometer (km)	=	1,000 meters (m)
1 meter	=	10 decimeters (dm)
	=	100 centimeters (cm)
	=	1,000 millimeters (mm)
1 decimeter	=	10 centimeters
1 centimeter	=	10 millimeters

Units of Area

1 square meter (m^2)	=	100 square decimeters (dm^2)
	=	10,000 square centimeters (cm^2)
1 square decimeter	=	100 square centimeters
1 are (a)	=	100 square meters
1 hectare (ha)	=	100 ares
1 square kilometer (km^2)	=	100 hectares

Units of Volume

1 cubic meter (m^3)	=	1,000 cubic decimeters (dm^3)
	=	1,000,000 cubic centimeters (cm^3)
1 cubic decimeter	=	1,000 cubic centimeters

Units of Capacity

1 kiloliter (kL)	=	1,000 liters (L)
1 liter	=	1,000 milliliters (mL)

Units of Mass

1 metric ton (t)	=	1,000 kilograms (kg)
1 kilogram	=	1,000 grams (g)
1 gram	=	1,000 milligrams (mg)

Units of Time

1 century	=	100 years
1 decade	=	10 years
1 year (yr)	=	12 months
	=	52 weeks (plus one or two days)
	=	365 days (366 days in a leap year)
1 month (mo)	=	28, 29, 30, or 31 days
1 week (wk)	=	7 days
1 day (d)	=	24 hours
1 hour (hr)	=	60 minutes
1 minute (min)	=	60 seconds (sec)

U.S. Customary System

Units of Length

1 mile (mi)	=	1,760 yards (yd)
	=	5,280 feet (ft)
1 yard	=	3 feet
	=	36 inches (in.)
1 foot	=	12 inches

Units of Area

1 square yard (yd^2)	=	9 square feet (ft^2)
	=	1,296 square inches ($in.^2$)
1 square foot	=	144 square inches
1 acre	=	43,560 square feet
1 square mile (mi^2)	=	640 acres

Units of Volume

1 cubic yard (yd^3)	=	27 cubic feet (ft^3)
1 cubic foot	=	1,728 cubic inches ($in.^3$)

Units of Capacity

1 gallon (gal)	=	4 quarts (qt)
1 quart	=	2 pints (pt)
1 pint	=	2 cups (c)
1 cup	=	8 fluid ounces (fl oz)
1 fluid ounce	=	2 tablespoons (tbs)
1 tablespoon	=	3 teaspoons (tsp)

Units of Weight

1 ton (T)	=	2,000 pounds (lb)
1 pound	=	16 ounces (oz)

System Equivalents

1 inch is about 2.5 cm (2.54).

1 kilometer is about 0.6 mile (0.621).

1 mile is about 1.6 kilometers (1.609).

1 meter is about 39 inches (39.37).

1 liter is about 1.1 quarts (1.057).

1 ounce is about 28 grams (28.350).

1 kilogram is about 2.2 pounds (2.205).

1 hectare is about 2.5 acres (2.47).

Rules for Order of Operations

1. Do operations within parentheses or other grouping symbols before doing anything else.
2. Calculate all exponents.
3. Multiply or divide in order from left to right.
4. Add or subtract in order from left to right.

Reference

Symbols

Symbol	Meaning
+	plus or positive
−	minus or negative
∗, ×	multiplied by
÷, /	divided by
=	is equal to
≠	is not equal to
<	is less than
>	is greater than
≤	is less than or equal to
≥	is greater than or equal to
x^n	nth power of x
$\sqrt{x}$	square root of x
%	percent
$a{:}b$, a/b, $\frac{a}{b}$	ratio of a to b or a divided by b or the fraction $\frac{a}{b}$
°	degree
(a,b)	ordered pair
$\overleftrightarrow{AS}$	line AS
$\overline{AS}$	line segment AS
$\overrightarrow{AS}$	ray AS
∟	right angle
⊥	is perpendicular to
∥	is parallel to
△ABC	triangle ABC
∠ABC	angle ABC
∠B	angle B

Place-Value Chart

trillions	100B	10B	billions	100M	10M	millions	hundred-thousands	ten-thousands	thousands	hundreds	tens	ones	.	tenths	hundredths	thousandths
1,000 billions			1,000 millions			1,000,000s	100,000s	10,000s	1,000s	100s	10s	1s	.	0.1s	0.01s	0.001s
10^{12}	10^{11}	10^{10}	10^9	10^8	10^7	10^6	10^5	10^4	10^3	10^2	10^1	10^0	.	10^{-1}	10^{-2}	10^{-3}

Probability Meter

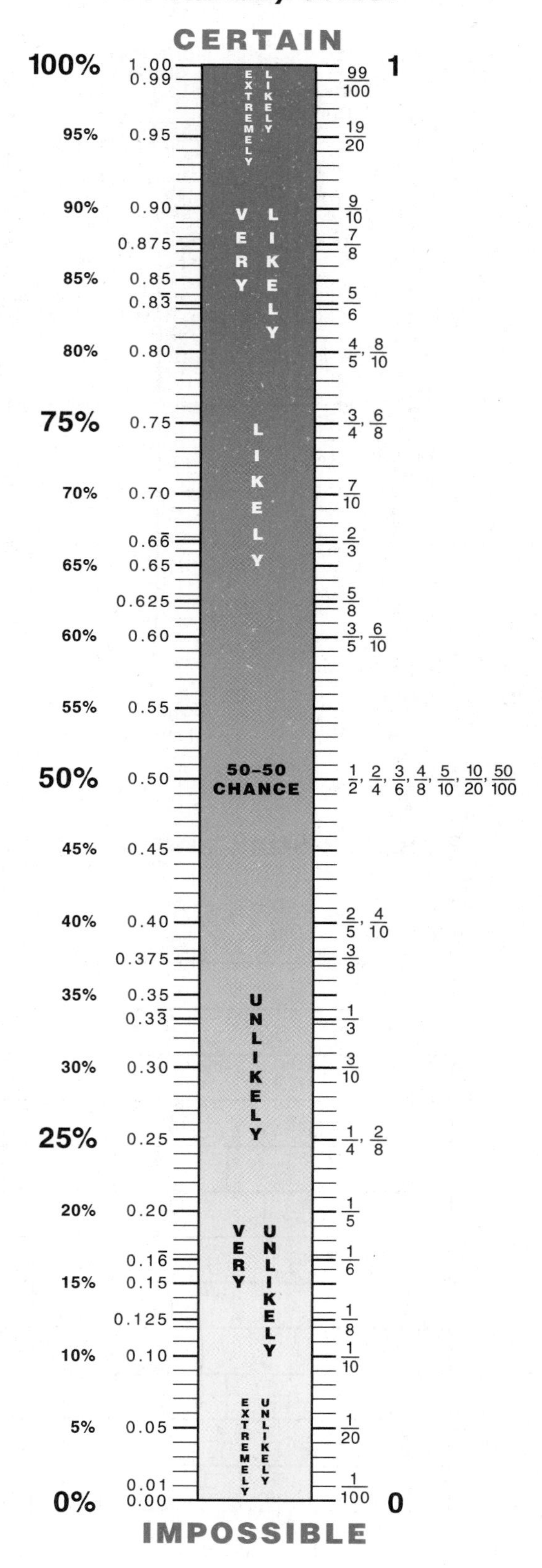

Date Time

Reference

Latitude and Longitude

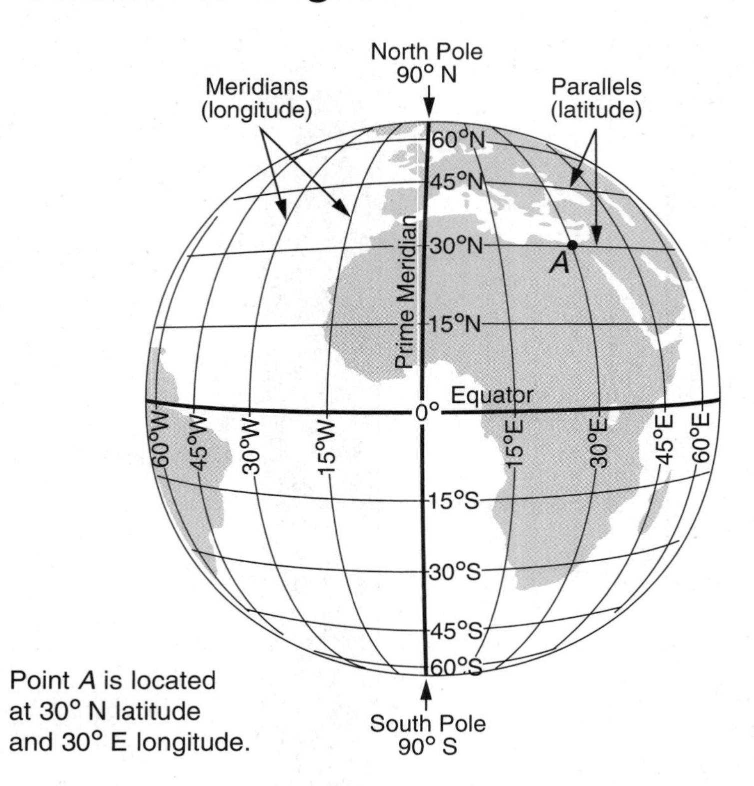

Point *A* is located at 30° N latitude and 30° E longitude.

Rational Numbers

Rule	Example
$\frac{a}{b} = \frac{n * a}{n * b}$	$\frac{2}{3} = \frac{4 * 2}{4 * 3} = \frac{8}{12}$
$\frac{a}{b} = \frac{a / n}{b / n}$	$\frac{8}{12} = \frac{8 / 4}{12 / 4} = \frac{2}{3}$
$\frac{a}{a} = a * \frac{1}{a} = 1$	$\frac{4}{4} = 4 * \frac{1}{4} = 1$
$\frac{a}{b} + \frac{c}{b} = \frac{a + c}{b}$	$\frac{3}{5} + \frac{1}{5} = \frac{3 + 1}{5} = \frac{4}{5}$
$\frac{a}{b} - \frac{c}{b} = \frac{a - c}{b}$	$\frac{3}{5} - \frac{1}{5} = \frac{3 - 1}{5} = \frac{2}{5}$
$\frac{a}{b} * \frac{c}{d} = \frac{a * c}{b * d}$	$\frac{1}{4} * \frac{2}{3} = \frac{1 * 2}{4 * 3} = \frac{2}{12}$

To compare, add, or subtract fractions:

1. Find a common denominator.
2. Rewrite fractions as equivalent fractions with the common denominator.
3. Compare, add, or subtract these fractions.

Fraction-Stick and Decimal Number-Line Chart

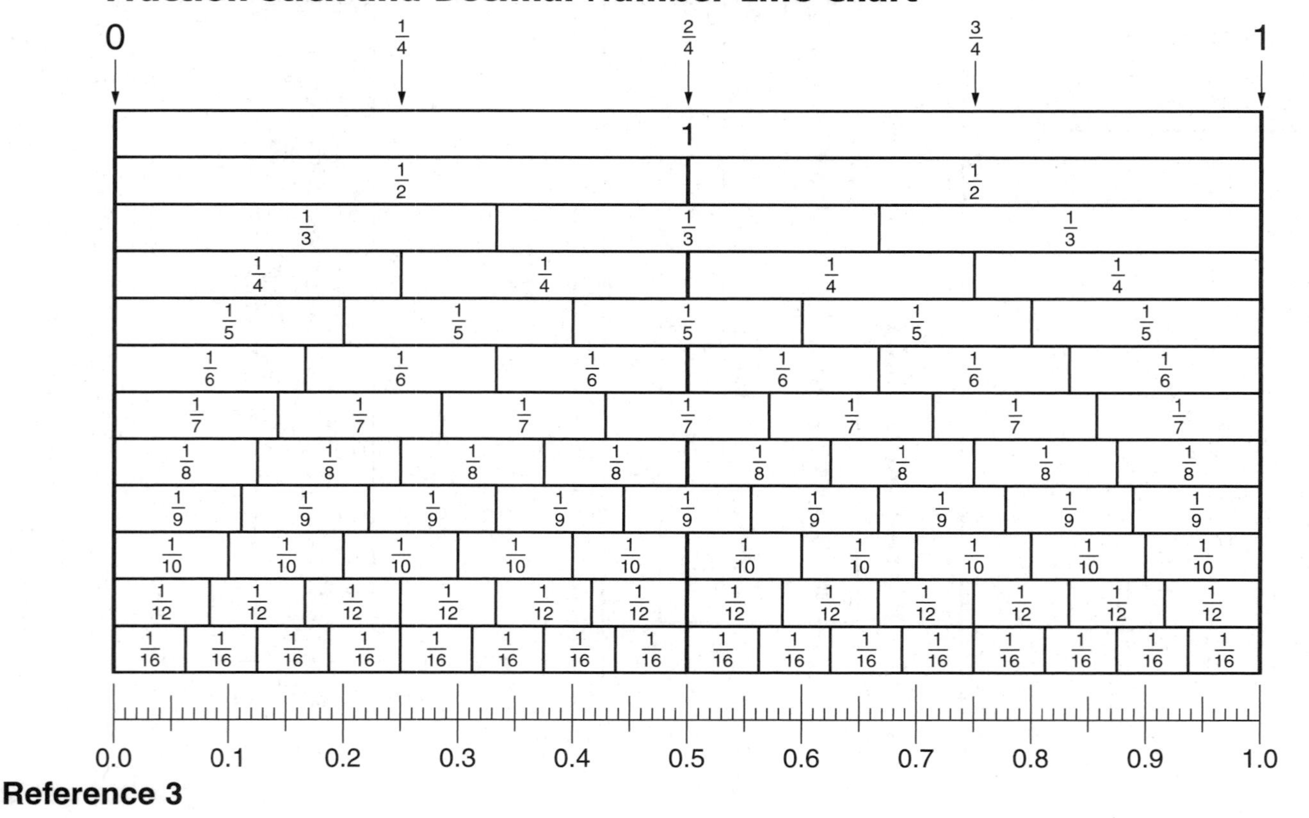

Date Time

Reference

Equivalent Fractions, Decimals, and Percents

$\frac{1}{2}$	$\frac{2}{4}$	$\frac{3}{6}$	$\frac{4}{8}$	$\frac{5}{10}$	$\frac{6}{12}$	$\frac{7}{14}$	$\frac{8}{16}$	$\frac{9}{18}$	$\frac{10}{20}$	$\frac{11}{22}$	$\frac{12}{24}$	$\frac{13}{26}$	$\frac{14}{28}$	$\frac{15}{30}$	0.5	50%
$\frac{1}{3}$	$\frac{2}{6}$	$\frac{3}{9}$	$\frac{4}{12}$	$\frac{5}{15}$	$\frac{6}{18}$	$\frac{7}{21}$	$\frac{8}{24}$	$\frac{9}{27}$	$\frac{10}{30}$	$\frac{11}{33}$	$\frac{12}{36}$	$\frac{13}{39}$	$\frac{14}{42}$	$\frac{15}{45}$	$0.\overline{3}$	$33\frac{1}{3}\%$
$\frac{2}{3}$	$\frac{4}{6}$	$\frac{6}{9}$	$\frac{8}{12}$	$\frac{10}{15}$	$\frac{12}{18}$	$\frac{14}{21}$	$\frac{16}{24}$	$\frac{18}{27}$	$\frac{20}{30}$	$\frac{22}{33}$	$\frac{24}{36}$	$\frac{26}{39}$	$\frac{28}{42}$	$\frac{30}{45}$	$0.\overline{6}$	$66\frac{2}{3}\%$
$\frac{1}{4}$	$\frac{2}{8}$	$\frac{3}{12}$	$\frac{4}{16}$	$\frac{5}{20}$	$\frac{6}{24}$	$\frac{7}{28}$	$\frac{8}{32}$	$\frac{9}{36}$	$\frac{10}{40}$	$\frac{11}{44}$	$\frac{12}{48}$	$\frac{13}{52}$	$\frac{14}{56}$	$\frac{15}{60}$	0.25	25%
$\frac{3}{4}$	$\frac{6}{8}$	$\frac{9}{12}$	$\frac{12}{16}$	$\frac{15}{20}$	$\frac{18}{24}$	$\frac{21}{28}$	$\frac{24}{32}$	$\frac{27}{36}$	$\frac{30}{40}$	$\frac{33}{44}$	$\frac{36}{48}$	$\frac{39}{52}$	$\frac{42}{56}$	$\frac{45}{60}$	0.75	75%
$\frac{1}{5}$	$\frac{2}{10}$	$\frac{3}{15}$	$\frac{4}{20}$	$\frac{5}{25}$	$\frac{6}{30}$	$\frac{7}{35}$	$\frac{8}{40}$	$\frac{9}{45}$	$\frac{10}{50}$	$\frac{11}{55}$	$\frac{12}{60}$	$\frac{13}{65}$	$\frac{14}{70}$	$\frac{15}{75}$	0.2	20%
$\frac{2}{5}$	$\frac{4}{10}$	$\frac{6}{15}$	$\frac{8}{20}$	$\frac{10}{25}$	$\frac{12}{30}$	$\frac{14}{35}$	$\frac{16}{40}$	$\frac{18}{45}$	$\frac{20}{50}$	$\frac{22}{55}$	$\frac{24}{60}$	$\frac{26}{65}$	$\frac{28}{70}$	$\frac{30}{75}$	0.4	40%
$\frac{3}{5}$	$\frac{6}{10}$	$\frac{9}{15}$	$\frac{12}{20}$	$\frac{15}{25}$	$\frac{18}{30}$	$\frac{21}{35}$	$\frac{24}{40}$	$\frac{27}{45}$	$\frac{30}{50}$	$\frac{33}{55}$	$\frac{36}{60}$	$\frac{39}{65}$	$\frac{42}{70}$	$\frac{45}{75}$	0.6	60%
$\frac{4}{5}$	$\frac{8}{10}$	$\frac{12}{15}$	$\frac{16}{20}$	$\frac{20}{25}$	$\frac{24}{30}$	$\frac{28}{35}$	$\frac{32}{40}$	$\frac{36}{45}$	$\frac{40}{50}$	$\frac{44}{55}$	$\frac{48}{60}$	$\frac{52}{65}$	$\frac{56}{70}$	$\frac{60}{75}$	0.8	80%
$\frac{1}{6}$	$\frac{2}{12}$	$\frac{3}{18}$	$\frac{4}{24}$	$\frac{5}{30}$	$\frac{6}{36}$	$\frac{7}{42}$	$\frac{8}{48}$	$\frac{9}{54}$	$\frac{10}{60}$	$\frac{11}{66}$	$\frac{12}{72}$	$\frac{13}{78}$	$\frac{14}{84}$	$\frac{15}{90}$	$0.1\overline{6}$	$16\frac{2}{3}\%$
$\frac{5}{6}$	$\frac{10}{12}$	$\frac{15}{18}$	$\frac{20}{24}$	$\frac{25}{30}$	$\frac{30}{36}$	$\frac{35}{42}$	$\frac{40}{48}$	$\frac{45}{54}$	$\frac{50}{60}$	$\frac{55}{66}$	$\frac{60}{72}$	$\frac{65}{78}$	$\frac{70}{84}$	$\frac{75}{90}$	$0.8\overline{3}$	$83\frac{1}{3}\%$
$\frac{1}{7}$	$\frac{2}{14}$	$\frac{3}{21}$	$\frac{4}{28}$	$\frac{5}{35}$	$\frac{6}{42}$	$\frac{7}{49}$	$\frac{8}{56}$	$\frac{9}{63}$	$\frac{10}{70}$	$\frac{11}{77}$	$\frac{12}{84}$	$\frac{13}{91}$	$\frac{14}{98}$	$\frac{15}{105}$	0.143	14.3%
$\frac{2}{7}$	$\frac{4}{14}$	$\frac{6}{21}$	$\frac{8}{28}$	$\frac{10}{35}$	$\frac{12}{42}$	$\frac{14}{49}$	$\frac{16}{56}$	$\frac{18}{63}$	$\frac{20}{70}$	$\frac{22}{77}$	$\frac{24}{84}$	$\frac{26}{91}$	$\frac{28}{98}$	$\frac{30}{105}$	0.286	28.6%
$\frac{3}{7}$	$\frac{6}{14}$	$\frac{9}{21}$	$\frac{12}{28}$	$\frac{15}{35}$	$\frac{18}{42}$	$\frac{21}{49}$	$\frac{24}{56}$	$\frac{27}{63}$	$\frac{30}{70}$	$\frac{33}{77}$	$\frac{36}{84}$	$\frac{39}{91}$	$\frac{42}{98}$	$\frac{45}{105}$	0.429	42.9%
$\frac{4}{7}$	$\frac{8}{14}$	$\frac{12}{21}$	$\frac{16}{28}$	$\frac{20}{35}$	$\frac{24}{42}$	$\frac{28}{49}$	$\frac{32}{56}$	$\frac{36}{63}$	$\frac{40}{70}$	$\frac{44}{77}$	$\frac{48}{84}$	$\frac{52}{91}$	$\frac{56}{98}$	$\frac{60}{105}$	0.571	57.1%
$\frac{5}{7}$	$\frac{10}{14}$	$\frac{15}{21}$	$\frac{20}{28}$	$\frac{25}{35}$	$\frac{30}{42}$	$\frac{35}{49}$	$\frac{40}{56}$	$\frac{45}{63}$	$\frac{50}{70}$	$\frac{55}{77}$	$\frac{60}{84}$	$\frac{65}{91}$	$\frac{70}{98}$	$\frac{75}{105}$	0.714	71.4%
$\frac{6}{7}$	$\frac{12}{14}$	$\frac{18}{21}$	$\frac{24}{28}$	$\frac{30}{35}$	$\frac{36}{42}$	$\frac{42}{49}$	$\frac{48}{56}$	$\frac{54}{63}$	$\frac{60}{70}$	$\frac{66}{77}$	$\frac{72}{84}$	$\frac{78}{91}$	$\frac{84}{98}$	$\frac{90}{105}$	0.857	85.7%
$\frac{1}{8}$	$\frac{2}{16}$	$\frac{3}{24}$	$\frac{4}{32}$	$\frac{5}{40}$	$\frac{6}{48}$	$\frac{7}{56}$	$\frac{8}{64}$	$\frac{9}{72}$	$\frac{10}{80}$	$\frac{11}{88}$	$\frac{12}{96}$	$\frac{13}{104}$	$\frac{14}{112}$	$\frac{15}{120}$	0.125	$12\frac{1}{2}\%$
$\frac{3}{8}$	$\frac{6}{16}$	$\frac{9}{24}$	$\frac{12}{32}$	$\frac{15}{40}$	$\frac{18}{48}$	$\frac{21}{56}$	$\frac{24}{64}$	$\frac{27}{72}$	$\frac{30}{80}$	$\frac{33}{88}$	$\frac{36}{96}$	$\frac{39}{104}$	$\frac{42}{112}$	$\frac{45}{120}$	0.375	$37\frac{1}{2}\%$
$\frac{5}{8}$	$\frac{10}{16}$	$\frac{15}{24}$	$\frac{20}{32}$	$\frac{25}{40}$	$\frac{30}{48}$	$\frac{35}{56}$	$\frac{40}{64}$	$\frac{45}{72}$	$\frac{50}{80}$	$\frac{55}{88}$	$\frac{60}{96}$	$\frac{65}{104}$	$\frac{70}{112}$	$\frac{75}{120}$	0.625	$62\frac{1}{2}\%$
$\frac{7}{8}$	$\frac{14}{16}$	$\frac{21}{24}$	$\frac{28}{32}$	$\frac{35}{40}$	$\frac{42}{48}$	$\frac{49}{56}$	$\frac{56}{64}$	$\frac{63}{72}$	$\frac{70}{80}$	$\frac{77}{88}$	$\frac{84}{96}$	$\frac{91}{104}$	$\frac{98}{112}$	$\frac{105}{120}$	0.875	$87\frac{1}{2}\%$
$\frac{1}{9}$	$\frac{2}{18}$	$\frac{3}{27}$	$\frac{4}{36}$	$\frac{5}{45}$	$\frac{6}{54}$	$\frac{7}{63}$	$\frac{8}{72}$	$\frac{9}{81}$	$\frac{10}{90}$	$\frac{11}{99}$	$\frac{12}{108}$	$\frac{13}{117}$	$\frac{14}{126}$	$\frac{15}{135}$	$0.\overline{1}$	$11\frac{1}{9}\%$
$\frac{2}{9}$	$\frac{4}{18}$	$\frac{6}{27}$	$\frac{8}{36}$	$\frac{10}{45}$	$\frac{12}{54}$	$\frac{14}{63}$	$\frac{16}{72}$	$\frac{18}{81}$	$\frac{20}{90}$	$\frac{22}{99}$	$\frac{24}{108}$	$\frac{26}{117}$	$\frac{28}{126}$	$\frac{30}{135}$	$0.\overline{2}$	$22\frac{2}{9}\%$
$\frac{4}{9}$	$\frac{8}{18}$	$\frac{12}{27}$	$\frac{16}{36}$	$\frac{20}{45}$	$\frac{24}{54}$	$\frac{28}{63}$	$\frac{32}{72}$	$\frac{36}{81}$	$\frac{40}{90}$	$\frac{44}{99}$	$\frac{48}{108}$	$\frac{52}{117}$	$\frac{56}{126}$	$\frac{60}{135}$	$0.\overline{4}$	$44\frac{4}{9}\%$
$\frac{5}{9}$	$\frac{10}{18}$	$\frac{15}{27}$	$\frac{20}{36}$	$\frac{25}{45}$	$\frac{30}{54}$	$\frac{35}{63}$	$\frac{40}{72}$	$\frac{45}{81}$	$\frac{50}{90}$	$\frac{55}{99}$	$\frac{60}{108}$	$\frac{65}{117}$	$\frac{70}{126}$	$\frac{75}{135}$	$0.\overline{5}$	$55\frac{5}{9}\%$
$\frac{7}{9}$	$\frac{14}{18}$	$\frac{21}{27}$	$\frac{28}{36}$	$\frac{35}{45}$	$\frac{42}{54}$	$\frac{49}{63}$	$\frac{56}{72}$	$\frac{63}{81}$	$\frac{70}{90}$	$\frac{77}{99}$	$\frac{84}{108}$	$\frac{91}{117}$	$\frac{98}{126}$	$\frac{105}{135}$	$0.\overline{7}$	$77\frac{7}{9}\%$
$\frac{8}{9}$	$\frac{16}{18}$	$\frac{24}{27}$	$\frac{32}{36}$	$\frac{40}{45}$	$\frac{48}{54}$	$\frac{56}{63}$	$\frac{64}{72}$	$\frac{72}{81}$	$\frac{80}{90}$	$\frac{88}{99}$	$\frac{96}{108}$	$\frac{104}{117}$	$\frac{112}{126}$	$\frac{120}{135}$	$0.\overline{8}$	$88\frac{8}{9}\%$

Note: The decimals for sevenths have been rounded to the nearest thousandth.

Date ______________________ Time ______________________

First to 100 Problem Cards

How many inches are in x feet? How many centimeters are in x meters? 1	How many quarts are in x gallons? 2	What is the smallest number of x's you can add to get a sum greater than 100? 3	Is $50 * x$ greater than 1,000? Is $\frac{x}{10}$ less than 1? 4
$\frac{1}{2}$ of $x = ?$ $\frac{1}{10}$ of $x = ?$ 5	$1 - x = ?$ $x + 998 = ?$ 6	If x people share 1,000 stamps equally, how many stamps will each person get? 7	What time will it be x minutes from now? What time was it x minutes ago? 8
It is 102 miles to your destination. You have gone x miles. How many miles are left? 9	What whole or mixed number equals x divided by 2? 10	Is x a prime or a composite number? Is x divisible by 2? 11	The time is 11:05 A.M. The train left x minutes ago. What time did the train leave? 12
Bill was born in 1939. Freddy was born the same day but x years later. In what year was Freddy born? 13	Which is larger: $2 * x$ or $x + 50$? 14	There are x rows of seats. There are 9 seats in each row. How many seats are there in all? 15	Sargon spent x cents on apples. If she paid with a $5 bill, how much change should she get? 16

Date Time

First to 100 Problem Cards *continued*

The temperature was 25°F. It dropped x degrees. What is the new temperature? 17	Each story in a building is 10 feet high. If the building has x stories, how tall is it? 18	Which is larger: $2 * x$ or $\frac{100}{x}$? 19	$20 * x = ?$ 20
Name all the whole-number factors of x. 21	Is x an even or an odd number? Is x divisible by 9? 22	Shalanda was born on a Tuesday. Linda was born x days later. On what day of the week was Linda born? 23	Will had a quarter plus x cents. How much money did he have in all? 24
Find the perimeter and area of this square. (x cm by x cm) 25	What is the median of these weights? 5 pounds, 21 pounds, x pounds. What is the range? 26	$x°$, ?° 27	$x^2 = ?$ 50% of $x^2 = ?$ 28
$(3x + 4) - 8 = ?$ 29	x out of 100 students voted for Ruby. Is this more than 25%, less than 25%, or exactly 25% of the students? 30	There are 200 students at Wilson School. x% speak Spanish. How many students speak Spanish? 31	People answered a survey question either Yes or No. x% answered Yes. What percent answered No? 32

Date _______________ Time _______________

Rotation Symmetry

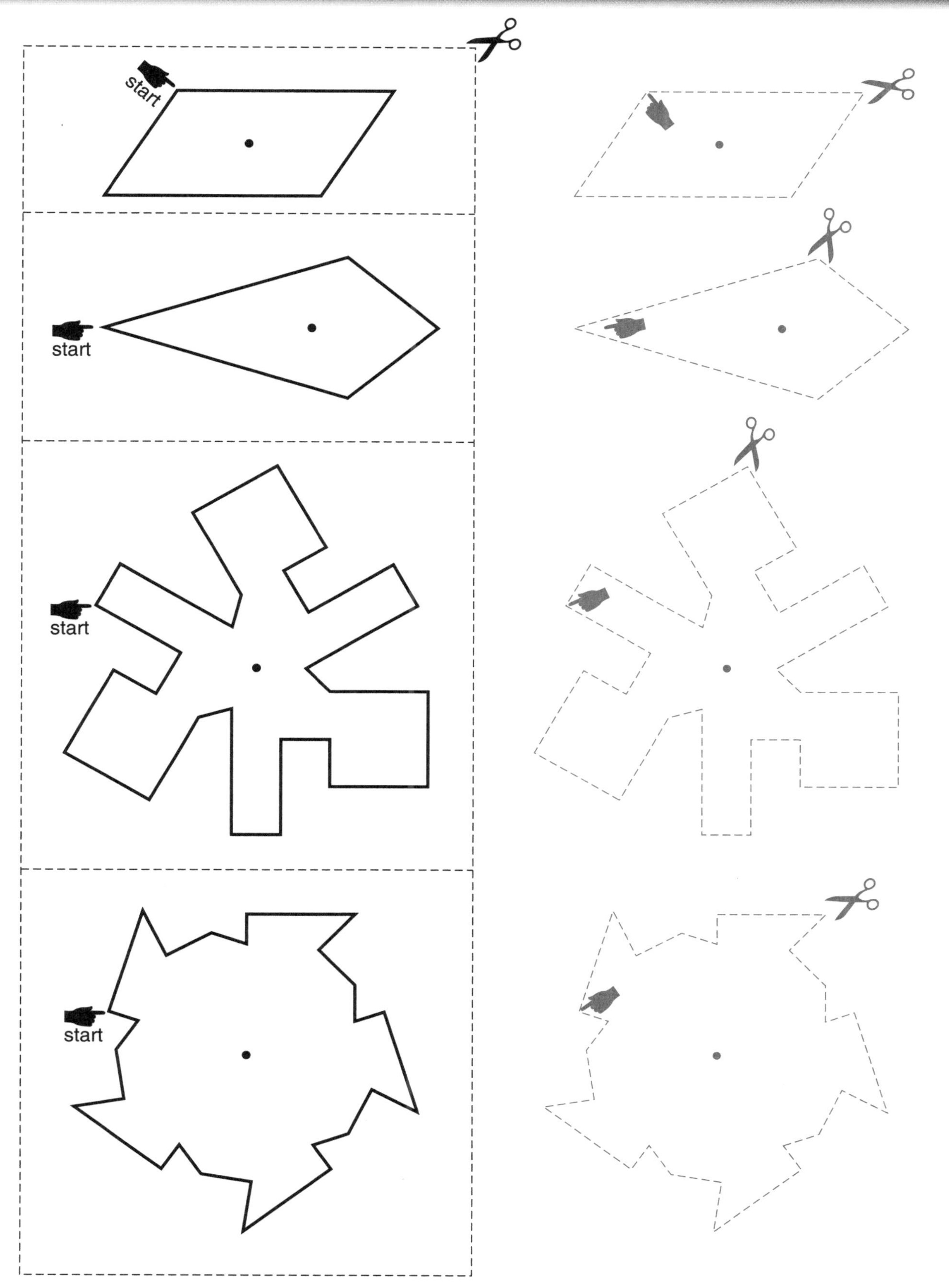